A-Level PHYSICS

Relativity and Quantum Physics

Roger Muncaster
B.Sc Ph.D
Formerly Head of Physics
Bury Metropolitan College
of Further Education

Stanley Thornes (Publishers) Ltd.

First published in 1995 by
Stanley Thornes (Publishers) Ltd
Ellenborough House
Wellington Street
Cheltenham
Gloucestershire GL50 1YD
England

A catalogue record for this book is available from the British Library.

ISBN 07487 1799 4

The front cover shows a molecular computer graphics image of the structure of one of the new generation of high-temperature superconductors: yttrium barium copper oxide ($YBa_2Cu_3O_{7-x}$). Copper atoms occur at the centre of layers of copper oxide square pyramids (green) and planes (blue). Oxygen atoms are in red, yttrium atoms in light blue and barium atoms in yellow.

Typeset by Tech-Set, Gateshead, Tyne & Wear.
Printed and bound at Scotprint, Musselburgh

Contents

Preface

This book is aimed specifically at Module Ph 9 (Turning Points in Physics) of the NEAB syllabuses in A-level and AS-level Physics. It also covers the section on Special Relativity in Module 4837 of the Cambridge A-level syllabuses in Physics, Single Award Science and Double Award Science.

The book contains many worked examples. Questions are included at relevant points in the text so that students can obtain an immediate test of their understanding of a topic. 'Consolidation' sections stress key points and in some cases present an overview of a topic in a manner which would not be possible in the main text. Definitions and fundamental points are highlighted – either by the use of screening or bold type. Questions, most of which are taken from past A-level papers, are included at the end of each of the nine chapters.

Acknowledgements

I wish to thank Veronica Hilton for her assistance with proof-reading and with the preparation of the index. I also wish to express my gratitude to the publishers, and to John Hepburn in particular, for their invaluable help throughout the preparation of the book. Many other people have helped me in a wide variety of ways – my sincere thanks to them.

I am indebted to the following examination boards for permission to use questions from their past examination papers:

Associated Examining Board [AEB]
University of Cambridge Local Examinations Syndicate [C], reproduced by permission of University of Cambridge Local Examinations Syndicate
Cambridge Local Examinations Syndicate, Overseas Examinations [C(O)]
Northern Examinations and Assessment Board (formerly the Joint Matriculation Board) [J]
Oxford and Cambridge Schools Examinations Board [O & C]
University of Oxford Delegacy of Local Examinations [O]
Southern Universities' Joint Board [S]
University of London Examinations and Assessment Council (formerly the University of London School Examinations Board) [L]
Welsh Joint Education Committee [W]

Thanks are also due to the following for providing photographs:

Ann Ronan Picture Library: p. 82(a)
Cavendish Laboratory, University of Cambridge: pp. 10, 58
Professor E.R. Huggins: p. 59
Science Photo Library: pp. 2, 21(b), 22, 27, 29, 37, 39, 54, 82(b), 87, 120(a); cover (Chemical Design Ltd); p. 131(a) (David Parker/IMI/University of Birmingham High TC Consortium); p. 21(a) (Dr Jeremy Burgess); p. 120(b) (J.-L. Charmet)
The Royal Society (by permission of the President and Council): p. 26
T. Matsumoto/SYGMA: p. 131(b)

R. MUNCASTER
Helmshore

1

THE ELECTRON AND ITS DISCOVERY

1.1 CATHODE RAYS

The conduction of electricity through gases at low pressures was investigated extensively for much of the latter half of the nineteenth century, following Johann Geissler's invention of an improved vacuum pump in 1855. These investigations led ultimately to the discovery of the electron.

The gas was normally contained in a **discharge tube** – a glass tube containing two metal electrodes and which has a small side arm so that it can be connected to a pump (Fig. 1.1). If there is a sufficiently high potential difference between the electrodes, the gas in the tube begins to glow when the pressure is reduced. The colour of the glow depends on the nature of the gas, its pressure and the PD across the tube. If the pressure is reduced below about 0.1 mmHg, the glow largely disappears and parts of the wall of the tube start to fluoresce. The colour of this fluorescence depends on the impurities present in the glass; it does not depend on the gas in the tube.

Fig. 1.1
A discharge tube

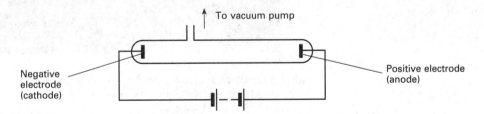

The fluorescence was first observed by Julius Plücker in 1859 and it soon became apparent that it was caused by invisible 'rays' of some kind emanating from the cathode. A German physicist, Eugen Goldstein, gave them the name of **cathode rays** in 1876. Plücker had found that the patches of fluorescence could be moved to different parts of the glass by placing a magnet close to the tube. Nowadays we would take this to be conclusive evidence that the cathode rays were charged particles of some sort, but things were not so clear-cut at the time. Though the consensus in England was that they were charged particles, that was not the case in Germany. Goldstein thought that the rays were a form of ultraviolet because ultraviolet was known to cause fluorescence. Heinrich Hertz was of the same opinion – he had been unable to deflect them in an electric field, which seemed to rule out the possibility of them being charged particles. He supposed instead that they were a form of ultraviolet that could be deflected by magnetic fields but not by electric fields. Hertz's assistant, Philipp Lenard, had shown that they could pass through thin metal foils <u>without puncturing them</u>. This was regarded as particularly strong evidence against the charged particle theory, for its proponents, notably Sir William Crookes and Cromwell Varley, imagined the particles to be the size of atoms and it was difficult to believe that anything so big would fail to puncture the foil.

Conclusive proof that cathode rays <u>are</u> charged particles, and furthermore that they are <u>negatively</u> charged, came from an experiment performed by Jean Baptiste Perrin in France in 1895. He used a discharge tube which had a small metal can mounted inside it. The can was connected to an electroscope and this acquired an increased negative charge when cathode rays entered the can.

Proof that cathode rays can be deflected by electric fields was first provided by J. J. Thomson in 1896. Thomson was also able to show that Hertz had failed because the pressure in his tube had been too high. (There would have been such a high concentration of ions present between the deflecting plates that the field between them would have been very severely reduced.) Finally, in 1897, Thomson showed that **cathode rays are what we now know as electrons** and is normally credited with having discovered the electron as a result. (See section 1.5.)

Joseph John Thomson
(1856 – 1940)

Cathode rays are created when fast-moving positive ions smash into the cathode of a discharge tube and dislodge electrons from it. The ions themselves are produced by electrons colliding with gas molecules, and are accelerated to high speeds by the PD across the tube. The glow observed at higher pressures is also the result of ionization by collision.

1.2 PROPERTIES OF CATHODE RAYS

(i) **They travel in straight lines.** An object (traditionally a metal plate in the shape of a Maltese cross) placed in the path of the 'rays' casts a well-defined shadow on a fluorescent screen.

(ii) **They can be deflected by both electric and magnetic fields**. The deflection is consistent with them being <u>negatively</u> charged particles.

(iii) **They cause fluorescence.**

(iv)   **They are emitted at 90° to the surface of the cathode.** The shadows of small objects produced by <u>plane</u> cathodes have sharp edges (i.e. they have no penumbra) which would not be the case if the rays were being emitted in many different directions. (This provided further evidence that cathode rays are <u>particles</u>. Light would be expected to be given off in all directions whereas charged particles would move parallel to the electric field lines, and these are perpendicular to the surface of the cathode.)

(v)   **They can be focused by using a concave cathode.** This follows from (iv) and is made use of in X-ray tubes.

(vi)   **They produce a heating effect.** A small piece of platinum placed in the path of the 'rays' has been observed to become white hot.

(vii) **They pass through thin metal foils without puncturing them.**

(viii) **They blacken photographic plates.**

(ix)   **Cathode rays in high-voltage discharge tubes cause X-rays to be emitted from the anode.** (Röntgen discovered X-rays in 1895 whilst using a discharge tube that had a platinum anode.)

1.3 WORK DONE ON AN ELECTRON BY A PD

Whenever an electron moves through a PD its kinetic energy changes by an amount equal to the work done by the PD. It follows that the kinetic energy, $\frac{1}{2}mv^2$, of an electron accelerated from rest through PD, V, is given by

$$\frac{1}{2}mv^2 = eV$$

where

$e = $ the charge on the electron ($= 1.6 \times 10^{-19}$ C).

QUESTIONS 1A

In the questions that follow, mass of electron $= 9.1 \times 10^{-31}$ kg, charge on electron $= 1.6 \times 10^{-19}$ C.

1. Calculate the speed of an electron which has been accelerated from rest through a PD of 2.0 V.

2. An electron moves from A to B. Its speed at A is 1.0×10^6 m s^{-1}. Calculate its speed on reaching B, **(a)** if B is 2.0 V positive relative to A, **(b)** if B is 2.0 V negative relative to A.

1.4 DEFLECTION OF ELECTRONS

In an Electric Field

It follows from the definition of electric field intensity that an electron in a field of intensity E is subject to a force F, given by

$$F = eE$$

where

$e = $ the charge on the electron ($= 1.6 \times 10^{-19}$ C).

Since electrons are <u>negatively</u> charged, the force is in the <u>opposite</u> direction to the field.

In the case of a <u>uniform</u> field (Fig. 1.2) due to two parallel plates with a separation d and across which there is a potential difference V, $E = V/d$ and therefore

$$F = e\frac{V}{d} \qquad \text{(for a \underline{uniform} field)}$$

Fig. 1.2
Force on an electron in a
uniform electric field

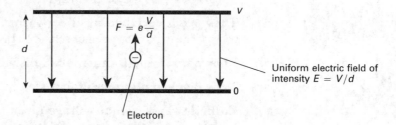

Consider a beam of electrons, each moving with velocity v, entering a <u>uniform</u> electric field of intensity E which is perpendicular to their direction of motion (Fig. 1.3). Once in the field each electron is subject to a force eE in the positive direction of the y-axis, and therefore by Newton's second law acquires an acceleration eE/m in this direction, where m is the mass of the electron. At the instant an electron enters the field its y-component of velocity is zero, and therefore after it has spent a time t in the field it will have undergone a vertical displacement, y, given by $s = ut + \frac{1}{2}at^2$ as

$$y = 0 + \frac{1}{2}\left(\frac{eE}{m}\right)t^2$$

i.e. $\qquad y = \frac{1}{2}\left(\frac{eE}{m}\right)t^2$ $\hfill$ [1.1]

Fig. 1.3
Deflection of an electron
beam in an electric field

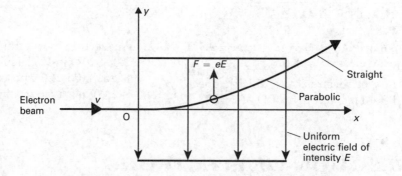

The electron's x-component of velocity is unaffected by the field and therefore x, its horizontal displacement from O, is given by

$$x = vt \hfill [1.2]$$

Eliminating t between equations [1.1] and [1.2] gives

$$y = \left(\frac{eE}{2mv^2}\right)x^2$$

This is the equation of a parabola, i.e. **the path of the electron whilst in the field is parabolic**. Once the electron has left the field its path is linear. Since the electron has gained a y-component of velocity whilst in the field and there has been no change in its x-component, its kinetic energy has increased.

The situation is analogous to that of a particle in a uniform gravitational field. In each case the magnitude and direction of the force are constant.

In a Magnetic Field

An electron moving with velocity v at right angles to a magnetic field of flux density B experiences a force F, which is given by

$$F = Bev$$

The force is perpendicular to both the field direction and the velocity, and its direction is given by Fleming's left-hand rule.

Consider an electron moving with velocity v into a uniform magnetic field of flux density B which is at right angles to its direction of motion (Fig. 1.4).

Fig. 1.4
Deflection of an electron beam in a magnetic field

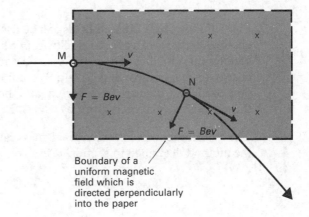

Boundary of a uniform magnetic field which is directed perpendicularly into the paper

On entering the field at M the electron feels a force F as shown and is deflected. The force is at right angles to the direction of motion of the electron and can neither speed it up nor slow it down. When the electron reaches some other point, N, the magnitude of the force acting on it is the same as it was at M (since none of B, e and v has changed) but the direction of the force is different. Thus, the force is perpendicular to the direction of motion at all times and has a constant magnitude, and therefore the electron travels with constant speed along a circular arc. Because the force is always at right angles to the velocity it does no work on the electron and therefore **the magnetic field does not change the kinetic energy of the electron**.

If the electron moves along an arc of radius r, then its centripetal acceleration v^2/r is given by Newton's second law as

$$Bev = \frac{mv^2}{r}$$

$$\therefore \quad r = \frac{mv}{Be}$$

Crossed Fields

If a uniform electric field and a uniform magnetic field are perpendicular to each other in such a way that they produce deflections in opposite senses, they are known as **crossed fields**. If the forces exerted by each field are of the same size, then

$$Bev = eE$$

i.e. $v = E/B$ [1.3]

Fig. 1.5
Principle of a velocity
selector

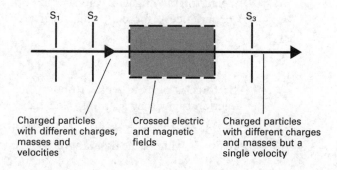

Charged particles
with different charges,
masses and
velocities

Crossed electric
and magnetic
fields

Charged particles
with different charges
and masses but a
single velocity

Fig. 1.5 shows how crossed fields can be used as a **velocity selector**, i.e. to select charged particles of a single velocity from a beam containing particles with a range of different velocities. The particles need not be electrons, and may have a range of charge-to-mass ratios. Slits S_1 and S_2 confine the particles to a narrow beam. The only particles which are undeflected, and therefore which emerge from slit S_3, are those whose velocity v is given by $v = E/B$.

The determination of the charge-to-mass ratio (specific charge) of the electron by the method described in section 1.5 makes use of crossed fields.

EXAMPLE 1.1

Refer to Fig. 1.6. A beam of electrons is accelerated through a PD of 500 V and then enters a uniform electric field of strength $3.00 \times 10^3 \, \text{V m}^{-1}$ created by two parallel plates each of length $2.00 \times 10^{-2} \, \text{m}$. Calculate: (a) the speed, v, of the electrons as they enter the field, (b) the time, t, that each electron spends in the field, (c) the angle, θ, through which the electrons have been deflected by the time they emerge from the field. (Specific charge (e/m) for electron $= 1.76 \times 10^{11} \, \text{C kg}^{-1}$.)

Fig. 1.6
Diagram for Example 1.1

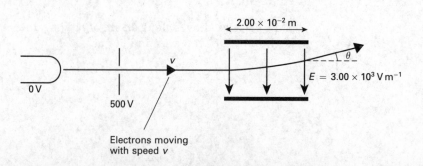

2.00×10^{-2} m

θ

$E = 3.00 \times 10^3 \, \text{V m}^{-1}$

0 V

500 V

Electrons moving
with speed v

Solution

(a) The kinetic energy gained by an electron is equal to the work done by the PD. Therefore

$$\tfrac{1}{2}mv^2 = eV$$

i.e. $v = \sqrt{\dfrac{2eV}{m}}$

$$= \sqrt{2 \times 1.76 \times 10^{11} \times 500} = 1.327 \times 10^7$$

i.e. Speed on entering field $= 1.33 \times 10^7 \, \mathrm{m\,s^{-1}}$

(b) The horizontal component of velocity is unaffected by the field between the plates and is therefore constant. It follows that t is given by

$$t = \frac{2.00 \times 10^{-2}}{1.327 \times 10^7} = 1.507 \times 10^{-9}$$

i.e. Time between plates $= 1.51 \times 10^{-9} \, \mathrm{s}$

(c) The electric field, E, exerts a force, F, on each electron where

$$F = eE$$

By Newton's second law this gives each electron an acceleration, a, towards the top of the page, where

$$a = \frac{F}{m} = \frac{eE}{m}$$

On emerging from the field the electron will have gained a vertical component of velocity, v_y, given by $v = u + at$ as

$$v_y = 0 + \left(\frac{eE}{m}\right)t$$

i.e. $v_y = \left(\dfrac{e}{m}\right) \times E \times t$

$$= 1.76 \times 10^{11} \times 3.00 \times 10^3 \times 1.507 \times 10^{-9} = 7.957 \times 10^5$$

Since $\tan \theta = v_y/v$

$$\tan \theta = \frac{7.957 \times 10^5}{1.327 \times 10^7} = 5.996 \times 10^{-2}$$

$$\therefore \quad \theta = 3.4°$$

EXAMPLE 1.2

Suppose that in an arrangement of the type described in Example 1.1, particles of charge, Q, and mass, M, are accelerated by a PD, V, and then enter a field of strength, E, between plates of length, d. Obtain an expression for the angle, θ, through which the particles will have been deflected by the time they leave the plates.

Solution

The speed, v, on entering the field between the plates is given by

$$\tfrac{1}{2}Mv^2 = QV \qquad \therefore \qquad v = \sqrt{\frac{2QV}{M}}$$

The time, t, between the plates is given by

$$t = \frac{d}{v}$$

The vertical acceleration, a, is given by

$$a = \frac{QE}{M}$$

The vertical component of velocity, v_y, is given by $v = u + at$ as

$$v_y = \left(\frac{QE}{M}\right)\frac{d}{v}$$

$$\therefore \qquad \tan\theta = \frac{v_y}{v} = \frac{QEd}{Mv^2}$$

Substituting for v gives

$$\tan\theta = \frac{QEd}{M}\frac{M}{2QV} = \frac{Ed}{2V}$$

i.e.

$$\theta = \tan^{-1}\left(\frac{Ed}{2V}\right)$$

Note that θ depends only on E, d and V – it does not depend on either Q or M. This result is interesting in itself, but it also serves to illustrate that we could have obtained the answer to Example 1.1 even if we had not been given the value of the specific charge of the particles involved. It further illustrates the effort that can be saved by not putting in numerical values until it is absolutely necessary!

EXAMPLE 1.3

An electron moving horizontally at $2.0 \times 10^5 \, \text{m s}^{-1}$ enters a uniform electric field which is directed vertically downwards and has a strength of $90 \, \text{V m}^{-1}$. The electron leaves the field $3.0 \times 10^{-8} \, \text{s}$ later with speed v. Find v.

(Mass of electron $= 9.1 \times 10^{-31} \, \text{kg}$, charge on electron $= 1.6 \times 10^{-19} \, \text{C}$.)

Solution

When the electron leaves the field its vertical component of velocity, v_y, is given by $v = u + at$ as

$$v_y = 0 + \left(\frac{eE}{m}\right)t$$

$$= \left(\frac{1.6 \times 10^{-19} \times 90}{9.1 \times 10^{-31}}\right) \times 3.0 \times 10^{-8} = 4.75 \times 10^5 \, \text{m s}^{-1}$$

The electron's horizontal component of velocity, v_x, has the constant value of $2.0 \times 10^5 \, \text{m s}^{-1}$, and therefore since

$$v^2 = v_x^2 + v_y^2$$
$$v^2 = (2.0 \times 10^5)^2 + (4.75 \times 10^5)^2$$

i.e. $v = 5.2 \times 10^5 \, \text{m s}^{-1}$

QUESTIONS 1B

1. An electron is moving in a circular path at $3.0 \times 10^6 \, \text{m s}^{-1}$ in a uniform magnetic field of flux density $2.0 \times 10^{-4} \, \text{T}$. Find the radius of the path. (Mass of electron $= 9.1 \times 10^{-31} \, \text{kg}$, charge on electron $= 1.6 \times 10^{-19} \, \text{C}$.)

2. An electron moving horizontally at $2.0 \times 10^6 \, \text{m s}^{-1}$ enters a uniform electric field of $300 \, \text{V m}^{-1}$ which is directed vertically downwards. The electron travels a horizontal distance of 10 cm whilst in the field. Find (a) the time for which the electron is in the field, (b) its vertical component of acceleration in the field,

(c) the amount by which it is displaced vertically, and (d) the speed with which it emerges from the field. (Charge on electron $= 1.6 \times 10^{-19} \, \text{C}$, mass of electron $= 9.1 \times 10^{-31} \, \text{kg}$.)

3. An electron is accelerated from rest through a PD of 60 V. It then enters a uniform magnetic field of flux density $4.0 \times 10^{-4} \, \text{T}$ and starts to travel along a circular path. Find the radius of the path. (Charge on electron $= 1.6 \times 10^{-19} \, \text{C}$, mass of electron $= 9.1 \times 10^{-31} \, \text{kg}$.)

1.5 DETERMINATION OF THE SPECIFIC CHARGE (*e/m*) OF THE ELECTRON AND ITS DISCOVERY

One form of apparatus used to determine *e/m* is shown in Fig. 1.7. Electrons which have been emitted thermionically (see section 1.8) by the filament are accelerated towards the cylindrical anode and pass through it. Two small holes on the axis of the anode confine the electrons to a narrow beam. When both fields (*E* and *B*) are zero the electrons reach the screen at X and produce fluorescence there.

Fig. 1.7
Apparatus to determine
e/m for the electron

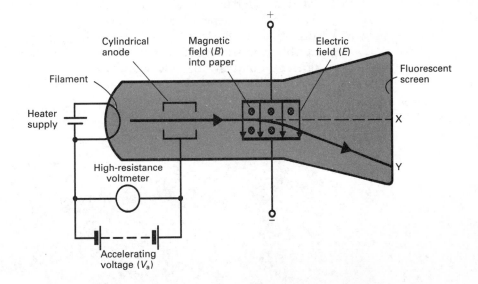

If the velocity of the electrons on emerging from the anode is v, then

$$eV_a = \tfrac{1}{2}mv^2$$

i.e. $$\frac{e}{m} = \frac{v^2}{2V_a}$$ [1.4]

where V_a = the accelerating voltage.

The position of X is noted and the magnetic field is switched on, deflecting the beam to Y. The electrons can be brought back to X by using the electric field, because it is arranged so that it exerts a force in the same region as the magnetic field but in the opposite direction. The electric field is switched on and is adjusted until the beam is again at X. The forces being exerted by each field must now be of equal size, and therefore

$$Bev = eE$$

i.e. $$v = E/B$$

Substituting for v in equation [1.4] gives

$$\frac{e}{m} = \frac{E^2}{2V_a B^2}$$

The value of E is found from $E = V/d$,

where V = the PD between the deflecting plates, and

d = their (known) separation.

The magnetic field is normally produced by a pair of Helmholtz coils, in which case the value of B can be found from the current through them, their radius and the number of turns. The value of V_a is read off directly from the voltmeter.

Discovery of the Electron

In 1897 J. J. Thomson used this method (but with a cold cathode rather than a heated filament) to determine the specific charge of **cathode rays** – the 'rays' which move from cathode to anode when an electric discharge is passed through a gas at low pressure. (The 'rays' were already known to be particles and there was evidence to suggest that in any particular discharge they all had the same specific charge, but there was no knowledge of the size of the charge, or of whether it was different under different conditions.) Thomson found that the value of the specific charge did not depend on the nature of the gas nor on the electrode material, suggesting that the 'rays' were composed of previously unknown particles which are a basic constituent of all matter. The idea that the particles were fundamentally different from any particle known at the time was strongly supported by

J. J. Thomson's e/m
tube for cathode rays

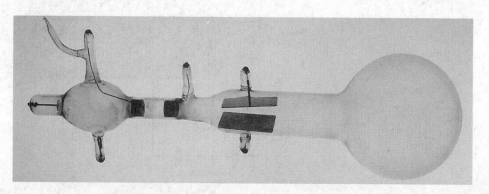

Thomson's value for their specific charge – of the order of a thousand times that of the hydrogen ion as found from experiments in electrolysis, and this was the highest of any known particle. The relatively high value of the specific charge of the cathode ray particles could be due to them having a large charge, or a small mass, or both. Since it was known that the particles could pass through thin metal foils without puncturing them, it seemed likely that the mass would be small and therefore Thomson assumed that they had the same charge as a monovalent ion, in which case they must be particles of very small mass. There is now no doubt that Thomson's assumption was correct. The particles whose specific charge he had measured are what are now called electrons, and in view of this, Thomson is normally credited with having discovered the electron.

1.6 DETERMINATION OF THE SPECIFIC CHARGE (e/m) OF THE ELECTRON BY USING A FINE-BEAM TUBE

The fine-beam tube (Fig. 1.8) is a glass bulb containing hydrogen at low pressure. Electrons are produced by an electron-gun arrangement at one side of the tube. The electrons collide with hydrogen atoms causing the atoms to emit light and so reveal the path of the electrons. A pair of Helmholtz coils provides a uniform magnetic field (directed perpendicularly out of the paper) which, provided it is sufficiently strong, deflects the electrons so that they travel in a complete circle. This circular path shows up as a luminous ring.

Fig. 1.8
The fine-beam tube

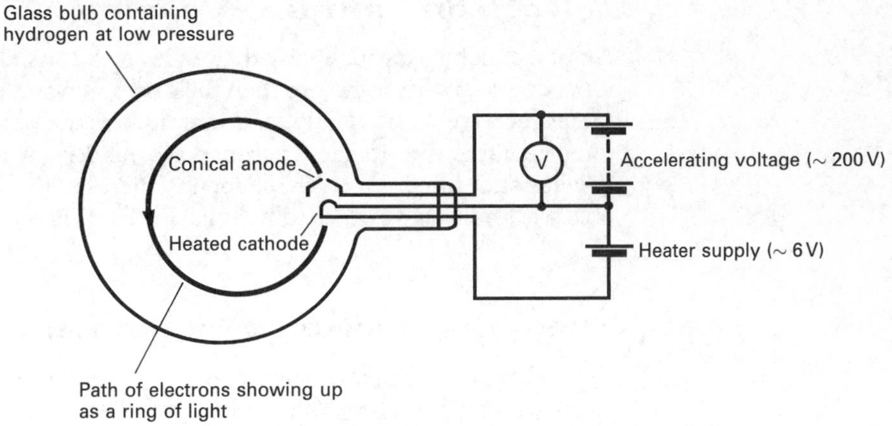

Glass bulb containing hydrogen at low pressure

Conical anode

Heated cathode

Path of electrons showing up as a ring of light

V

Accelerating voltage ($\sim$ 200 V)

Heater supply ($\sim$ 6 V)

If r = the radius of the electron path,

v = the velocity of the electrons on leaving the electron gun, and

B = the magnetic flux density,

then the force on the electron is Bev, the centripetal acceleration is v^2/r, and therefore by Newton's second law

$$Bev = mv^2/r \qquad\qquad [1.5]$$

Also, if the accelerating voltage (i.e. the PD between the anode and cathode of the electron gun) is V, then

$$\tfrac{1}{2}mv^2 = eV \qquad\qquad [1.6]$$

because eV is the work done by the accelerating PD, and $\tfrac{1}{2}mv^2$ is the kinetic energy gained by the electrons as a result. By equation [1.5]

$$v = Br(e/m) \qquad\qquad\qquad\qquad [1.7]$$

By equation [1.6]

$$v^2 = 2V(e/m) \qquad\qquad\qquad\qquad [1.8]$$

By equations [1.7] and [1.8]

$$(Br)^2(e/m)^2 = 2V(e/m)$$

$$\therefore \quad \frac{e}{m} = \frac{2V}{B^2 r^2}$$

Hence e/m may be determined.

A high-resistance voltmeter is used to measure V, and B can be found from the current through the Helmholtz coils, their radius and number of turns. The diameter, and therefore the radius, of the electron path can be measured by placing a mirror with a scale on it behind the tube and lining up the luminous ring with its image.

1.7 MILLIKAN'S DETERMINATION OF e (1909)

A Note on Terminal Velocity

An object falling through air experiences a viscous drag. Initially the downward force due to gravity is greater than the drag force and the object accelerates. Drag forces increase with velocity, and therefore as the object accelerates, the upward directed force due to the drag increases and eventually becomes equal to the gravitational force. Once this happens there is no further acceleration and the object is said to have reached its **terminal velocity**.

Theory of Millikan's Experiment

The principle of Millikan's experiment is to measure the terminal velocity of a small, charged oil drop falling under gravity, and then to oppose its motion with an electric field in such a way that it remains stationary.

When there is no electric field the forces acting on the drop are as shown in Fig. 1.9(a). Once the drop has reached its terminal velocity it has no acceleration, and therefore

Fig. 1.9
Forces acting on an oil drop in Millikan's oil-drop experiment:
(a) without electric field,
(b) with electric field

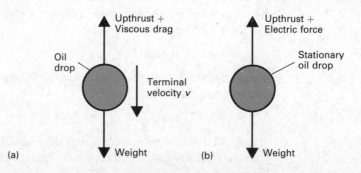

$$\text{Weight} = \text{Upthrust due to air} + \text{Viscous drag} \qquad [1.9]$$

But

$$\text{Weight} = \text{Volume of drop} \times \text{Density of oil} \times g$$

Therefore for a spherical drop

$$\text{Weight} = \tfrac{4}{3}\pi r^3 \rho_o g$$

where

$$r = \text{radius of drop}$$

$$\rho_o = \text{density of oil.}$$

Also

$$\text{Upthrust} = \text{Weight of air displaced by drop}$$

$$= \text{Volume of drop} \times \text{Density of air} \times g$$

i.e. $$\text{Upthrust} = \tfrac{4}{3}\pi r^3 \rho_a g$$

where

$$\rho_a = \text{density of air.}$$

By Stokes' law

$$\text{Viscous drag} = 6\pi r \eta v$$

where

$$\eta = \text{coefficient of viscosity of air.}$$

Substituting in equation [1.9] gives

$$\tfrac{4}{3}\pi r^3 \rho_o g = \tfrac{4}{3}\pi r^3 \rho_a g + 6\pi r \eta v \qquad [1.10]$$

When an electric field has been applied such that the drop is stationary, the forces acting on the drop are as shown in Fig. 1.9(b). The drop has no velocity and no acceleration and therefore

$$\text{Weight} = \text{Upthrust} + \text{Electric force}$$

i.e. $$\tfrac{4}{3}\pi r^3 \rho_o g = \tfrac{4}{3}\pi r^3 \rho_a g + QE \qquad [1.11]$$

where

$$Q = \text{the charge on the drop}$$

$$E = \text{the electric field strength.}$$

Subtracting equation [1.10] from equation [1.11] gives

$$0 = QE - 6\pi r \eta v$$

i.e. $$Q = \frac{6\pi r \eta v}{E} \qquad [1.12]$$

Millikan measured E and v and did a separate experiment to find η. He was not able to measure r directly, but by equation [1.10]

$$\tfrac{4}{3}\pi r^3 (\rho_o - \rho_a)g = 6\pi r \eta v$$

i.e. $$r = \left(\frac{9\eta v}{2(\rho_o - \rho_a)g} \right)^{1/2}$$

Substituting for r in equation [1.12] gives

$$Q = \frac{6\pi\eta}{E}\left(\frac{9\eta v}{2(\rho_o - \rho_a)g}\right)^{1/2} v \qquad\qquad [1.13]$$

Note The density of air at room temperature and pressure is less than one thousandth of that of oil, and except for very accurate calculations equation [1.13] can be replaced by

$$Q = \frac{6\pi\eta}{E}\left(\frac{9\eta v}{2\rho_o g}\right)^{1/2} v$$

This of course is the result that would have been obtained had the upthrust due to the displaced air been ignored in the first place.

Experimental Procedure

The apparatus is shown in schematic form in Fig. 1.10. A and B are two metal plates which are accurately parallel to each other. (In Millikan's apparatus the plates had a diameter of 20 cm and a separation of 1.5 cm.) An atomizer is used to create a fine mist of oil drops in the region of the small hole in the upper plate. The drops are charged, either positively or negatively, as a result of losing or gaining electrons through frictional effects on emerging from the atomizer. Some of the drops fall through the hole and are observed by reflected light through a low-power microscope. The eyepiece of the microscope incorporates a calibrated graticule so that the terminal velocity v of any particular oil drop can be determined by timing its fall through a known distance.

Fig. 1.10
Millikan's apparatus to
determine the charge on
the electron

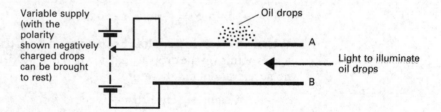

An electric field ($\sim 10^5\,\text{V m}^{-1}$) is applied at this stage and is adjusted so that the drop whose velocity has just been determined is held stationary. The strength E of this field is given by $E = V/d$, where V is the PD between the plates and d is their separation.

Millikan measured the charges on hundreds of drops. **The charges were always integral multiples of 1.6×10^{-19} C, and he concluded that electric charge can never exist in fractions of this amount and that the magnitude of the electronic charge e is 1.6×10^{-19} C,** i.e. Millikan established the **quantization of electric charge.**

Notes (i)   For some measurements Millikan used X-rays to ionize the air through which the drops fell. When this was done the speeds of some of the drops changed suddenly (though not all at the same time). Millikan interpreted this as being due to the drops colliding with ions and acquiring positive or negative charges according to whether they had collided with a positive ion or a negative ion. He found that the change in charge was always plus or minus a small whole-number multiple of e.

(ii) A constant-temperature enclosure surrounded Millikan's apparatus in order to eliminate convection currents. It also served to shield the apparatus from draughts.

(iii) Millikan used a low-vapour-pressure oil to reduce problems due to evaporation.

(iv) Millikan's actual technique was to cause the drops to move upwards under the influence of the electric field – we have considered the drops to be at rest in order to simplify the theory.

(v) Millikan's early results showed that it is necessary to use a modified form of Stokes' law to account for very small drops.

QUESTIONS 1C

1. In an experiment to measure the charge on the electron by Millikan's method, the charges on five drops were found to be:

   $69\,k$, $118\,k$, $35\,k$, $68\,k$ and $84\,k$

   where k is a constant. In terms of k, what is the charge on the electron?

2. An oil drop with a mass of 1.2×10^{-15} kg is held stationary between two parallel metal plates whose separation is 6.0 mm. The drop carries a charge of 4.8×10^{-19} C. Ignoring the upthrust due to air, calculate (a) the electric field strength between the plates, (b) the PD across the plates. ($g = 10\,\mathrm{m\,s^{-2}}$.)

1.8 THERMIONIC EMISSION

All metals contain some electrons which are free to move about within the lattice. Though the attractive forces exerted on these electrons by the atomic nuclei are not strong enough to bind them to particular atoms, they do prevent them from leaving the surface. When a metal is heated the energies of its electrons increase and some of them acquire sufficient energy to escape from the surface. The process is called **thermionic emission**. The rate at which electrons are emitted increases rapidly with temperature. It can be shown on theoretical grounds that if the electrons are drawn towards a positively charged electrode so that they constitute a current I, then

$$I = AT^2 e^{-W/kT} \qquad [1.14]$$

This is known as **Richardson's equation**; A and W are constants which are characteristic of the emitter, k is Boltzmann's constant and T is the temperature of the emitter in kelvins.

In equation [1.14] W is the minimum amount of energy which has to be supplied to the metal to remove an electron from the surface of the metal; it is called the **work function** of the metal. There is good agreement between work function values estimated on the basis of the thermionic effect (i.e. from equation [1.14]) and those estimated on the basis of the photoelectric effect (section 3.7) – see Table 1.1.

Table 1.1
Thermionic and photoelectric work functions compared

Metal	W/eV Thermionic	W/eV Photoelectric
Caesium	1.81	1.9
Tungsten	4.52	4.49

CONSOLIDATION

Cathode Rays

The 'rays' emitted by the cathode of a discharge tube containing a gas at low pressure when a current is passed through the gas. They are the result of positive ion bombardment of the cathode and cause the glass wall of the discharge tube to fluoresce. They are now known to be electrons.

Plücker showed that they could be deflected by magnetic fields which suggested that they were charged particles. Others thought they could not possibly be charged particles (i) because they could penetrate thin metal foils without puncturing them, and (ii) because Hertz had not been able to deflect them with an electric field. Perrin provided conclusive proof that they are negatively charged particles in 1895.

In 1897 Thomson measured the **specific charge** of cathode rays and found that: (i) it is independent of both the electrode material and the gas in the tube, and (ii) it is of the order of a thousand times greater than that of the hydrogen ion. He concluded that the cathode ray particles are a basic constituent of all matter – what are now called **electrons**.

Properties of Cathode Rays – see section 1.2.

An electron follows a parabolic path in a uniform electric field (except when it moves parallel to the field lines).

An electron moving in a plane which is at right angles to a uniform magnetic field follows a circular path.

An electric field changes the kinetic energy of an electron.

A magnetic field cannot change the kinetic energy of an electron (because the force it exerts is always at right angles to the motion and therefore does no work).

For an electron in an **electric field**

$$F = eE \qquad F = e\frac{V}{d} \text{ (uniform field)} \qquad eV = \tfrac{1}{2}mv^2$$

For an electron in a **magnetic field**

$$F = Bev \qquad Bev = \frac{mv^2}{r} \qquad \therefore r = \frac{mv}{Be}$$

Thermionic Emission

$$I = AT^2 e^{-W/kT} \qquad\qquad (\textbf{Richardson's equation})$$

QUESTIONS ON CHAPTER 1

1. Electrons projected along a horizontal path in a cathode ray tube may be deflected on their way to the screen by a horizontal magnetic field. Draw a diagram of this arrangement and show on your diagram the directions of the beam, the magnetic field and the deflection. For an electron in a vacuum tube, $e\Delta V = \tfrac{1}{2}mv^2$.

Explain what the term $e\Delta V$ corresponds to. In television sets, electrons are typically accelerated through 25 kV. Calculate the speed that they reach. (The electronic charge $= -1.6 \times 10^{-19}$ C, electronic mass $= 9.1 \times 10^{-31}$ kg.)

[L, '93]

2. (a) Electrons are accelerated from rest through a PD of 4000 V in an evacuated tube, then they enter a uniform magnetic field B of flux density 10^{-3} T which is at right angles to the electron beam as shown in the diagram below.

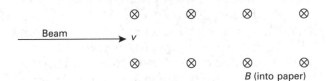

Beam v

B (into paper)

(i) Calculate the speed v of the electrons on entering the magnetic field.

(ii) Calculate the magnitude of the force experienced by an electron in the magnetic field.

(iii) Copy the above diagram and show the direction of this force by means of an arrow.

(b) Explain why the electrons move in a circular path and calculate the radius of this path. ($e = 1.6 \times 10^{-19}$ C, $m_e = 9.1 \times 10^{-31}$ kg.)

[W, '93]

3. (a) Figure 1 shows a heated tungsten filament which acts as an electron source in an evacuated cathode ray tube.

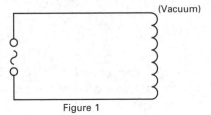

(Vacuum)

Figure 1

(i) Explain (with a diagram) how the electrons from the filament can be accelerated to produce a horizontal beam.

(ii) The specific charge of electrons is 1.8×10^{11} C kg^{-1}. Calculate the accelerating voltage required to accelerate the electrons in the beam to a maximum speed of 1.0×10^7 m s^{-1}.

(b) The beam of electrons from the filament passes through a vertical electric field (Figure 2).

(i) State how the velocities of the electrons will be affected.

(ii) Describe fully the direction of a suitable magnetic field that could

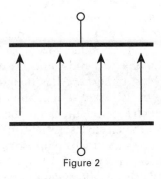

Figure 2

be superimposed on the electric field so that electrons with a single speed v would pass through the combined fields without deflection.

(iii) Derive a relation between the field strength E of the electric field, the flux density B of the magnetic field, and the electron speed v for this condition of zero deflection. [O, '93]

4. (a) An electron (mass m, charge e) travels with speed v in a circle of radius r in a plane perpendicular to a uniform magnetic field of flux density B.

(i) Write down an algebraic equation relating the centripetal and electromagnetic forces acting on the electron.

(ii) Hence show that the time for one orbit of the electron is given by the expression $T = 2\pi m/Be$.

(b) If the speed of the electron changed to $2v$, what effect, if any, would this change have on:

(i) the orbital radius r,

(ii) the orbital period T?

(c) Radio waves from outer space are used to obtain information about interstellar magnetic fields. These waves are produced by electrons moving in circular orbits. The radio wave frequency is the same as the electron orbital frequency.

(The mass of an electron is 9.1×10^{-31} kg, and its charge is -1.6×10^{-19} C.)

If waves of frequency 1.2 MHz are observed, calculate:

(i) the orbital period of the electrons;

(ii) the flux density of the magnetic field. [O, '93]

5. Describe, giving the theory and a labelled diagram of the apparatus, a method of determining e/m for the electron.

In an evacuated tube, electrons are accelerated through a potential difference of 500 V. Calculate their final speed, and consider whether this depends on the accelerating field being uniform. After this acceleration, the electrons pass through a uniform electric field which is perpendicular to the direction of travel of the electrons as they enter the field. This electric field is produced by applying a potential difference of 10 V to two parallel plates which are 0.06 m long and 0.02 m apart. Assume that the electric field is uniform and confined to the space between the two plates. Determine the angular deflection of the electron beam produced by the field.
(e/m for the electron $= 1.76 \times 10^{11}\,\mathrm{C\,kg^{-1}}$.)

[W]

6. Two parallel metal sheets of length 10 cm are separated by 20 mm in a vacuum. A narrow beam of electrons enters symmetrically between them as shown.

When a PD of 1000 V is applied between the

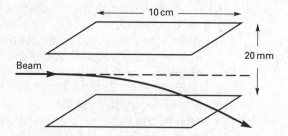

plates the electron beam just misses one of the plates as it emerges.

Calculate the speed of the electrons as they enter the gap. (Take the field between the plates to be uniform.)

($e/m = 1.8 \times 10^{11}\,\mathrm{C\,kg^{-1}}$.)   [W, '92]

7. A heated filament and an anode with a small hole in it are mounted in an evacuated glass tube so that a narrow beam of electrons emerges vertically upwards from the hole in the anode. A uniform magnetic field is applied so that the electrons describe a circular path in a vertical plane.

(a) Draw a diagram showing the path of the electrons and indicate the direction of the magnetic field which will cause the beam to curve in the direction you have shown. Explain why the path is circular.

(b) Derive an expression for the specific electronic charge (e/m) of the electrons in terms of the PD between the anode and

filament, V, the radius of the circular path, r, and the magnetic flux density, B.

(c) What value of B would be required to give a radius of the electron path of $2r$, assuming that V remains constant? If B is now held constant at its new value, what value of V will restore the beam to its former radius?

(d) Describe and account for the changes in **(i)** kinetic energy, **(ii)** momentum which an electron undergoes from the instant it leaves the heated filament with negligible velocity until it has completed a full circle in the magnetic field.   [J]

8. The diagram shows a type of cathode ray tube containing a small quantity of gas. Electrons from a hot cathode emerge from a small hole in a conical shaped anode, and the path subsequently followed is made visible by the gas in the tube.

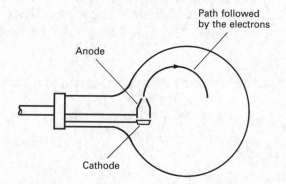

(a) The accelerating voltage is 5.0 kV. Calculate the speed of the electrons as they emerge from the anode.

(b) The apparatus is situated in a uniform magnetic field acting into the plane of the diagram. Explain why the path followed by the beam is circular. Calculate the radius of the circular path for a flux density of 2.0×10^{-3} T.

(c) Suggest a possible process by which the gas in the tube might make the path of the beam visible.

(Specific charge of an electron $= 1.8 \times 10^{11}\,\mathrm{C\,kg^{-1}}$.)   [AEB, '87]

9. **(a)** Explain what is meant by quantization of charge.

(b) A cloud of oil droplets is formed between two horizontal parallel metal plates. Explain the following observations:

(i) In the absence of an electric field between the plates, all the oil droplets fall slowly at uniform speeds.

(ii) On applying a vertical electric field, some droplet speeds are unaltered, some are increased downwards, whereas some droplets move upwards. [W, '90]

10. Robert Millikan observed the motion of charged oil droplets between two oppositely charged horizontal plates 5.0 mm apart. In such an experiment, a charged droplet of mass 9.80×10^{-16} kg was balanced by the upward electric force created by applying a PD of 300 V between the plates.

(a) If the top plate was positive with respect to the bottom plate, what was the sign of the charge on the droplet?

(b) Calculate
(i) the electric field strength between the plates,
(ii) the charge on the droplet.

(c) What conclusions did Millikan come to after measuring the charge on many individual droplets?   ($g = 9.8 \, \text{m s}^{-2}$.)
[J (specimen), '96]

11. (a) A charged oil drop falls at constant speed in the Millikan oil-drop experiment when there is no PD between the plates. Explain this.

(b) Such an oil drop, of mass 4.0×10^{-15} kg, is held stationary when an electric field is applied between the two horizontal plates. If the drop carries 6 electric charges each of value 1.6×10^{-19} C, calculate the value of the electric field strength.
(Assume $g = 9.8 \, \text{m s}^{-2}$.)   [L]

12. (a) A charged drop of oil is held stationary in the space between two metal plates across which an electric field is applied. Explain with the aid of a diagram how the forces acting enable the drop to remain stationary.

(b) In an investigation using the above arrangement the voltage across the plates is recorded. The drop is then given a different charge and the voltage adjusted until the drop is stationary again. This procedure is repeated several times and the voltage recorded on each occasion. The following voltage values were obtained:

142, 425, 569, 709, 999.

Discuss whether these results support the notion that the charge on the drop is a multiple of some fundamental value. [S]

13. In a measurement of the electron charge by Millikan's method, a potential difference of 1.5 kV can be applied between horizontal parallel metal plates 12 mm apart. With the field switched off, a drop of oil of mass 10^{-14} kg is observed to fall with constant velocity $400 \, \mu\text{m s}^{-1}$. When the field is switched on, the drop rises with constant velocity $80 \, \mu\text{m s}^{-1}$. How many electron charges are there on the drop? (You may assume that the air resistance is proportional to the velocity of the drop, and that air buoyancy may be neglected.)
(The electronic charge $= 1.6 \times 10^{-19}$ C, the acceleration due to gravity $= 10 \, \text{m s}^{-2}$.) [S]

14. (a) In an experiment to attempt to confirm Millikan's conclusion that the electron carries a discrete charge, a charged oil droplet of known mass was held stationary between a pair of parallel horizontal plates by an electric field. The PD producing the field was measured. The charge on the drop was changed and the new voltage to maintain equilibrium was measured. The experiment was repeated and the following results were obtained.

Measurement	PD required/V
1	225
2	110
3	150
4	50

(i) Draw a diagram which shows the forces acting on the oil drop when it is held stationary in the field. State the origins of the forces.

(ii) Suggest a means by which the charge on the oil drop could be changed.

(iii) Explain clearly how these measurements suggest that Millikan's conclusion was correct.

(iv) The separation of the parallel plates was 0.020 m. Making the assumption that the oil drop in the first set of observations carried an excess of two electrons, calculate the force acting on the oil drop when the PD was switched off.
The charge on an electron, $e = -1.6 \times 10^{-19}$ C.

(b) In order to determine the mass of an electron the specific charge of an electron has first to be found.

(i) State what is meant by the term *specific charge* and state the unit in which it is measured.

(ii) Describe an experiment to determine a value for the specific charge. Your account should include

—a diagram of the apparatus showing essential circuitry

—the experimental procedure stating clearly the measurements you would make

—an explanation of the theory of the method.   [AEB, '90]

15. In Millikan's experiment an oil drop of mass 1.92×10^{-14} kg is stationary in the space between the two horizontal plates which are 2.00×10^{-2} m apart, the upper plate being earthed and the lower one at a potential of -6000 V. State, with the reason, the sign of the electric charge on the drop. Neglecting the buoyancy of the air, calculate the magnitude of the charge.

With no change in the potentials of the plates, the drop suddenly moves upwards and attains a uniform velocity. Explain why **(a)** the drop moves, **(b)** the velocity becomes uniform.

(The acceleration due to gravity $= 10 \, \mathrm{m\,s^{-2}}$.)
                                                                            [J]

16. A small oil drop, carrying a negative electric charge, is falling in air with a uniform speed of 8.00×10^{-5} m s^{-1} between two horizontal parallel plates. The upper plate is maintained at a positive potential relative to the lower one. Draw a diagram showing all the forces acting on the drop, stating the cause of each force.

Use the following data to determine the charge on the oil drop.

Radius of drop	$= 1.60 \times 10^{-6}$ m.
Density of oil	$= 800 \, \mathrm{kg\,m^{-3}}$.
Density of air	$= 1.30 \, \mathrm{kg\,m^{-3}}$.
Viscosity of air	$= 1.80 \times 10^{-5} \, \mathrm{N\,s\,m^{-2}}$.
Distance between plates	$= 1.00 \times 10^{-2}$ m.
PD between plates	$= 2.00 \times 10^{3}$ V.
Acceleration of free fall, g	$= 10 \, \mathrm{m\,s^{-2}}$. [L]

17. Describe, with the necessary theory, some experiment by which the charge on the electron has been determined.

An oil drop, of mass 3.2×10^{-15} kg, falls vertically with uniform velocity through the air between vertical parallel plates 3 cm apart. When a potential difference of 2000 V is applied between the plates, the drop moves with uniform velocity at an angle of $45°$ to the vertical. Calculate the charge on the drop.

The path of the drop suddenly changes, becoming inclined at $18°26'$ to the vertical; later, the path changes again and becomes inclined at $33°42'$ to the vertical. Estimate from these data the elementary unit of charge (electron charge).   [S]

2

LIGHT: WAVES OR PARTICLES?

2.1 NEWTON'S CORPUSCULAR THEORY. HUYGENS' WAVE THEORY

By the end of the seventeenth century there were two conflicting theories as to the nature of light. Newton, like most scientists at the time, regarded rays of light as streams of particles (corpuscles) that are emitted by luminous objects and which travel in straight lines until they reach the boundary of a different medium. The other theory was (principally) the work of Christiaan Huygens, an eminent Dutch scientist, who proposed, in 1678, that light is a kind of wave motion. Huygens knew that beams of light can cross each other and emerge from the crossing point without having been affected in any way. This, of course, is characteristic of wave behaviour – particles would be expected to collide with each other and scatter in many different directions.

(a) Isaac Newton
(1642–1727)
(b) Christiaan Huygens
(1629–95)

For more than a century the corpuscular theory was much more widely accepted than the wave theory. There were three main reasons for this.

(i) **Newton's powerful reputation.**

(ii) **An opaque object in the path of a beam of light casts a sharp shadow.**
Particles travelling in straight lines would be expected to do just this. On the

other hand, if light were a wave motion, it ought to bend round an obstacle in its path, as other wave motions were known to do, in which case it would not produce sharp shadows.

(iii)   **Light can travel through a vacuum**. Whilst there was obviously no difficulty with the idea of particles travelling through a vacuum, there was with waves – no known wave motion could!

Opinion began to swing towards the wave theory in 1801 when Thomas Young showed that the colours produced in thin films illuminated by white light can be explained on the basis of the interference of light waves. These effects had long been known but had never received a satisfactory explanation on the basis of the corpuscular theory. Young went so far as to calculate the wavelengths of visible light, in some cases using measurements that had been made by Newton himself.

Despite this success, the wave theory was still not generally accepted. Young failed to communicate his ideas in a manner that was acceptable to the scientific community of his day and therefore few took his work seriously. Furthermore, objectors continued to argue that if light were a wave motion, it would not produce sharp shadows. However, in the years that followed, Fresnel developed a rigorous mathematical theory backed up by a number of carefully conceived experiments, and by about 1820 there were few who doubted the validity of the wave theory. Fresnel showed that the ability of light to produce sharp shadows is due to its extremely short wavelength. Waves do not bend appreciably around objects that are large in comparison with their wavelength, and therefore any bending that occurs with light normally goes unnoticed.

Augustin Jean Fresnel
(1788–1827)

The final blow to the corpuscular theory came in 1850 when Jean Foucault measured the velocity of light in air and in water. In order to account for refraction, **the wave theory requires that light travels more quickly in air than in water; the corpuscular theory requires the opposite**. Foucault's result confirmed the wave theory prediction and is generally regarded as being the conclusive evidence against the corpuscular theory.

Notes   (i)   The waves envisioned by Huygens were waves in the sense that they transfer energy rather than matter, but he imagined them as a series of random pulses in a medium called the ether (section 6.1); there was no regularity associated with them, i.e. no wavelength or frequency.

(ii)   Newton was not totally uncompromising in his support for the corpuscular theory. On balance, though, he favoured it over the wave theory because he could not come to terms with the idea that light could propagate as a wave and still produce sharp shadows.

(iii)   In a report published posthumously in 1665, Francesco Grimaldi described experiments in which he had placed a narrow aperture in the path of a beam of light. He found that light spread into regions that would have been in shadow if light travelled only in straight lines. Both Newton and Grimaldi interpreted this as being due to some sort of refraction caused by particles of light passing close to the edge of the aperture – mistakenly, neither regarded it as evidence in support of the wave theory.

2.2 REFLECTION, REFRACTION AND DISPERSION ACCORDING TO NEWTON'S CORPUSCULAR THEORY

Reflection

Newton supposed that particles of light were acted on by a repulsive force when they approached close to a reflecting surface (Fig. 2.1). This reduced their component of velocity perpendicular to the surface to zero and then increased it to its original value but in the opposite direction. The force was perpendicular to the surface and so had no effect on the parallel component of velocity. The particles would therefore follow a curved path whilst in the region where the force was operative. Since the magnitude of neither component changed (overall), the particles would leave the surface with the same speed as they had approached and would do so at the same angle, i.e. $i = r$. Thus the corpuscular theory accounts for the second law of reflection.

Fig. 2.1
Reflection on the basis of
Newton's corpuscular
theory

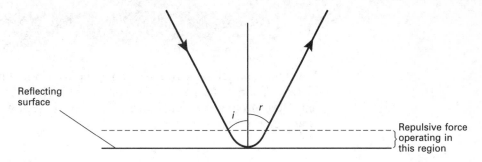

Reflecting
surface

Repulsive force
operating in
this region

The theory also accounts for the first law. The incident ray is in the plane of the paper, and since there is no force perpendicular to this plane, the reflected ray must also be in the plane of the paper, i.e. in the same plane as the incident ray.

Refraction

Consider a ray of light incident in air on the boundary with an optically more dense material such as water (Fig. 2.2). In order to account for the refraction that occurs, Newton supposed that whilst they were in a narrow region that existed on each side of the boundary, the particles were acted on by a force which increased their normal components of velocity. The force was perpendicular to the boundary and therefore had no effect on the components of velocity parallel to the surface. It

Fig. 2.2
Refraction on the basis of
Newton's corpuscular
theory

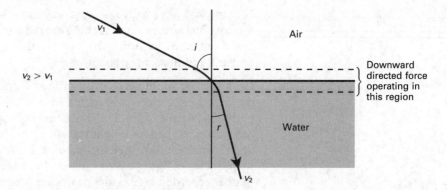

follows that the ray would bend <u>towards</u> the normal, as required, and that there would be an overall increase in velocity. Thus **the corpuscular theory predicts that light speeds up on entering an optically more dense medium**.

Since the parallel component of velocity is unchanged, we see from Fig. 2.2 that

$$v_1 \sin i = v_2 \sin r$$

$$\therefore \quad \frac{\sin i}{\sin r} = \frac{v_2}{v_1} \tag{2.1}$$

From Snell's law

$$\frac{\sin i}{\sin r} = n \tag{2.2}$$

where n is the refractive index of water relative to air. Combining equations [2.1] and [2.2] gives

$$n = \frac{v_2}{v_1}$$

i.e. $$\text{Refractive index of water relative to air} = \frac{\text{Velocity of light in water}}{\text{Velocity of light in air}} \quad \left(\begin{matrix}\text{On the basis of the}\\ \text{corpuscular theory}\end{matrix}\right)$$

Reflection and refraction often occur at the same time. In order to account for this, Newton assumed the particles had phases which he called **'fits of easy reflection'** and **'fits of easy transmission'**, and were reflected or refracted according to which phase they were in.

Dispersion

Newton explained **dispersion** (the splitting of white light into its component colours, by a prism, for example) by assuming that white light is a mixture of particles of different colours. It appears white simply because that is the effect that is given when all the colours are together.

When light is refracted the various colours are subject to different forces; violet particles feel the strongest force and red the weakest. They are therefore refracted along different paths and so reveal their separate identities. The essential point is that the colours are not <u>created</u> by the refraction, they are already present in the white light. A particle of any particular colour always has that colour. Experiment confirms this. If white light is passed through a prism and a single colour is then isolated and passed through a second prism, no further splitting occurs.

2.3 YOUNG'S DOUBLE-SLIT EXPERIMENT

The experiment provides a clear demonstration of the interference of light waves. We shall give only a brief description of (a modern version of) the experiment; more detailed accounts can be found elsewhere*. The reader should be familiar with the phenomenon of optical interference and of the need for coherence before continuing.

The experimental arrangement is shown schematically in Fig. 2.3. S, S_1 and S_2 are narrow slits which are parallel to each other; the separation of S_1 and S_2 is typically 1 mm. Because S is narrow it diffracts the light that falls on it and so illuminates both S_1 and S_2. Diffraction also takes place at S_1 and S_2, and interference occurs in the region where the light from S_1 overlaps that from S_2. Because S is narrow, the light which emerges from S_1 comes from the same point as that which emerges from S_2 and therefore S_1 and S_2 act as coherent sources.

Fig. 2.3
The double-slit
experiment

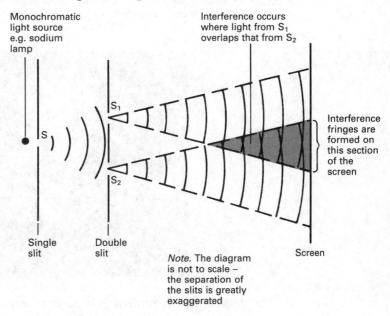

A series of alternately bright and dark bands (**interference fringes**) which are equally spaced and parallel to the slits can be observed on a screen placed <u>anywhere</u> in the region of overlap.

The fringes are accounted for on the basis of the wave theory. The centres of the bright bands are at places where the waves from S_1 and S_2 arrive at the screen in phase with each other; the centres of the dark bands are where the two waves are totally out of phase with each other.

The corpuscular theory can offer no convincing explanation for the fringes because it is impossible to imagine how two (or more) particles could interact in such a way as to produce darkness. The experiment therefore provides strong evidence in support of the wave theory of light.

Historical Note

Young's original experiment is commonly supposed to have been performed in either 1801 or 1802. However, the first reference to it appears to be that in a lecture which Young gave in 1807[†] and this gives no indication of when he actually carried

---

*See, for example, R. Muncaster, *A-Level Physics* (Stanley Thornes).
[†] Thomas Young: A Course of Lectures on natural Philosophy and the Mechanical Arts (1807), Volume 1, pp. 457–71.

out the experiment. Young provides only a sketchy account of the experimental arrangement; it is not clear, for example, whether he used slits or pinholes – probably both! In discussing the conditions under which two sources of light can be caused to produce observable interference, Young says:

> ... the simplest case appears to be, when a beam of homogeneous [monochromatic] light falls on a screen in which there are two very small holes or slits, which may be considered as centres of divergence, from whence the light is diffracted in every direction. In this case, when the two newly formed beams are received on a surface placed so as to intercept them, their light is divided by dark stripes into portions nearly equal, but becoming wider as the surface is more remote from the apertures, so as to subtend very nearly equal angles from the apertures at all distances, and wider also in the same proportion as the apertures are closer to each other. The middle of the two portions is always light, and the bright stripes on each side are at such distances, that the light, coming to them from one of the apertures, must have passed through a longer space than that which comes from the other, by an interval which is equal to the breadth of one, two, three, or more of the supposed undulations, while the intervening dark spaces correspond to a difference of half a supposed undulation, of one and a half, of two and a half, or more.

Thomas Young
(1773–1829)

The double-slit experiment (whenever it was first performed) clearly provided strong evidence for the validity of the wave theory. Furthermore, it was the first successful experiment designed specifically for this purpose. However, it should be understood that it was Young's explanation of the colours in thin films (see section 2.1) that created the revised interest in the wave theory in the early nineteenth century, not the double-slit experiment.

2.4 ELECTROMAGNETIC WAVES

In the process known as electromagnetic induction, discovered by Faraday in 1831, an EMF is induced in a conductor which is situated in a changing magnetic field. There is an electric field associated with this induced EMF and therefore **a changing magnetic field creates an electric field**. It turns out that **the two fields are at right angles to each other** and that an electric field is created whenever a magnetic field changes, even in the absence of a conductor.

In 1864 Maxwell suggested that the converse effect might also occur, i.e. that a magnetic field would be created by a changing electric field. If so, it follows that when a magnetic field is changing in such a way as to produce a changing electric field, then this in turn will produce a changing magnetic field which will produce a changing electric field, and so on. Maxwell predicted that an oscillating charge

James Clerk Maxwell
(1831–1879)

would produce just such an effect and that as a result a disturbance consisting of time-varying electric and magnetic fields would propagate through space in the form of a transverse wave – an **electromagnetic wave**.

The varying electric and magnetic fields are perpendicular to each other, have the same frequency (the frequency of the wave) and are in phase with each other. Maxwell was able to calculate the speed of the waves. The value turned out to be very close to the known value of the speed of light in vacuum and was regarded as strong evidence that light was a form of electromagnetic wave motion, and therefore that electromagnetic waves did in fact exist. On the basis of Maxwell's theory the speed, c, of electromagnetic waves in a vacuum is given by

$$c = \frac{1}{\sqrt{\mu_0 \varepsilon_0}}$$

[2.3]

where μ_0 and ε_0 are constants called the permeability of free space (vacuum) and the permittivity of free space respectively. It is now generally accepted that light is an electromagnetic wave motion and that c in equation [2.3] can be regarded as the speed of light in vacuum. Nowadays the ampere is defined in such a way that $\mu_0 = 4\pi \times 10^{-7}$ henrys per metre ($H\,m^{-1}$) exactly and the metre is defined such that $c = 299\,792\,458\,m\,s^{-1}$. Substituting these values in equation [2.3] gives $\varepsilon_0 = 8.854$ farads per metre ($F\,m^{-1}$).

Summary

(i)   An oscillating electric charge generates **electromagnetic waves**.

(ii)   The waves are transverse and consist of oscillating electric and magnetic fields which are perpendicular to each other and to the direction of travel of the wave.

(iii)   The oscillating electric and magnetic fields have the same frequency as each other – the frequency of the wave.

(iv)   Their speed, c, is given by $c = 1/\sqrt{\mu_0 \varepsilon_0}$ and is equal to the speed of light in vacuum.

2.5 HERTZ'S DISCOVERY OF RADIO WAVES (1887)

The fact that the speed of light appeared to be the same as that predicted for electromagnetic waves did not constitute actual proof of their existence. The first conclusive evidence came from an experiment performed by Heinrich Hertz in 1887.

Hertz used a **spark-gap transmitter** (Fig 2.4(a)) to produce what are now known as **radio waves**. The metal plates were charged to a high PD by use of an **induction coil** – a device which produces <u>pulses</u> of high voltage. Whenever the PD across the

Fig. 2.4
(a) Spark-gap transmitter,
(b) Wire loop receiver

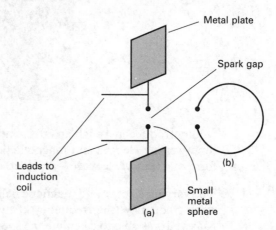

air gap between the spheres was high enough to break down the insulating properties of the air a spark would jump across the gap and discharge the plates. The discharge takes the form of a high-frequency damped oscillation (Fig. 2.5).

Fig. 2.5
Discharge of spark-gap transmitter

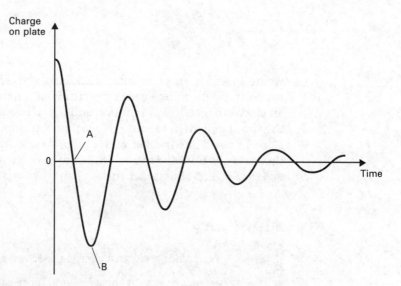

The circuit behaves in this way because it is effectively a series combination of capacitance, inductance and resistance. (The plates and air gap constitute a capacitor and the rods between the plates and spheres have a small inductance. The air gap is the main source of resistance.) Because there is inductance in the circuit, the discharge current builds up slowly and does not reach its maximum value until the plates are fully discharged (at A). The inductance then delays the current falling to zero and because the current is still flowing in the same direction, the plates charge up again with the opposite polarity (at B). The plates now start to discharge and the oscillations persist until the energy has been dissipated, primarily as heat in the air gap. At this stage a second high-voltage pulse from the induction coil starts the whole process again.

Thus a high-frequency alternating current flows in the circuit whenever the plates discharge. This means that charge is oscillating at high frequency, and that electromagnetic waves should be generated in accordance with Maxwell's theory (section 2.4).

Heinrich Rudolf Hertz
(1857–94)

Hertz detected the waves using a wire loop which had a small gap in it. The time-varying magnetic field of the electromagnetic wave induced an EMF in the loop and this caused sparks to pass across the gap in it. Hertz was able to show that the waves were polarized in the way that Maxwell had predicted – the sparks stopped when the loop was turned through 90° about the direction of propagation of the wave. He also demonstrated that the waves were transmitted by insulators but reflected by metals.

The Velocity of the Waves

Hertz assumed that there was a strong spark only when the resonant frequency of his detector (Fig. 2.4(b)) was equal to the frequency of the waves. This allowed him to work out their frequency (f) by calculating the resonant frequency of the detector from its dimensions.

The wavelength was determined by reflecting the waves from a metal sheet in such a way that stationary waves were created. He could locate the positions of the nodes by moving the detector along the stationary wave pattern and noting where it failed to spark. Since the separation of adjacent nodes is half a wavelength, the wavelength (λ) could be measured.

The velocity (v) of the waves is given by $v = f\lambda$ and was very close to the value predicted by Maxwell – conclusive proof that Hertz had detected the waves whose existence Maxwell had predicted some thirteen years earlier.

2.6 MICHELSON'S MEASUREMENT OF THE SPEED OF LIGHT

There have been many determinations of the velocity of light. The first of these was that of a Danish astronomer, Römer, in 1676. (Galileo had made an unsuccessful attempt in 1600.) An extremely accurate, and historically important, measurement was made by Michelson in 1926. The method involved a rotating octagonal steel prism (Fig. 2.6).

When the prism is stationary the light follows the path shown and an image of the source can be seen through the telescope. If the prism is rotated slowly, the image disappears because either face X is not in a suitable position to direct the outgoing beam to C, or face Y is unable to send the incoming beam to the telescope. However, if the speed of the prism is increased so that it turns through exactly one-eighth of a revolution in the same time that it takes light to travel from X to Y, then an image of the source is seen through the telescope. Michelson adjusted the speed of rotation until he was able to observe a stationary image of the source. This occurred when the prism was rotating at $530 \, \text{rev s}^{-1}$. The experiment was carried out on Mt. Wilson (USA) and the concave reflector (C) was on another mountain 35 km away.

Fig. 2.6
Michelson's method for determining the velocity of light. (Note that the optical system used by Michelson was more complex than that shown here.)

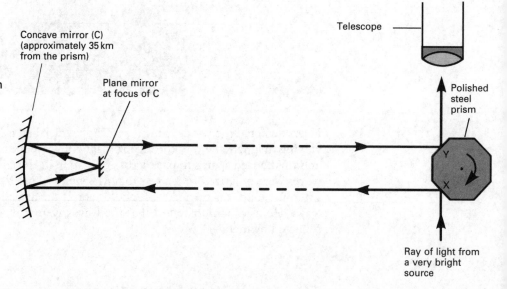

The figures given here are approximate but can be used to indicate the procedure adopted by Michelson. Thus,

$$\text{Speed of light } (c) = \frac{\text{Light path}}{\text{Time taken}}$$

$$= \frac{2 \times 35 \times 10^3}{(1/8)(1/530)}$$

$$\approx 3 \times 10^8 \, \text{m s}^{-1}$$

Michelson made a correction to take account of the fact that the light was travelling through air (rather than vacuum) and obtained a final value which was accurate to about one part in 10^5.

CONSOLIDATION

Corpuscular Theory and Wave Theory

At the time of Newton and Huygens (late seventeenth century) there were two theories of the nature of light.

The fact that two beams of light could pass through each other without either being affected suggested that light was a wave motion (Huygens' view).

The fact that light could travel through a vacuum and that it (apparently) produced sharp shadows suggested that it was a particle motion (Newton's view).

Both theories could account for reflection and refraction.

The majority of scientists favoured the corpuscular (particle) theory.

In 1801 Thomas Young explained the colours in thin films illuminated by white light on the basis of the interference of light waves. This was strong evidence in favour of the wave theory but Young's ideas did not receive general acceptance until Fresnel had developed a rigorous mathematical theory and had explained how light could produce sharp shadows.

In 1850 Foucault provided final confirmation of the wave theory when he found by experiment that light travels more slowly in water than in air.

Electromagnetic Waves

In 1864 Maxwell predicted that an electromagnetic disturbance could travel through a vacuum in the form of a transverse wave moving at the speed of light.

Hertz demonstrated the existence of these waves in 1887.

$$c = 1/\sqrt{\mu_0 \varepsilon_o}$$

QUESTIONS ON CHAPTER 2

1. (a) At the end of the seventeenth century Newton's corpuscular theory of light was much more widely accepted than Huygens' wave theory. Explain why.

   (b) By the end of the nineteenth century the corpuscular theory had been totally discarded in favour of the wave theory. Explain why.

2. (a) Thomas Young demonstrated interference of light using a double-slits arrangement and a candle flame which is a light source that emits a continuous spectrum of light. The same phenomenon can be demonstrated using monochromatic light from a laser. Give two reasons why interference fringes produced by a candle flame are much more difficult to observe than fringes produced by a laser beam.

   (b) Explain why Newton's corpuscular theory of light cannot explain the formation of interference fringes whereas Huygens' wave theory can.

   (c) Why was the wave theory of light not accepted until the nineteenth century?

   [J (specimen), '96]

3. Give a brief account of Hertz's experiment that confirmed the existence of electromagnetic waves.

4. Give an account of the main features of an electromagnetic wave.

5. The following expression for the speed c of light in free space was proposed by Maxwell in 1864:

$$c = \frac{1}{(\mu_0 \varepsilon_0)^{1/2}}$$

where μ_0 and ε_0 are respectively the magnetic permeability and electric permittivity of free space.

(a) Show that the unit of $(\mu_0 \varepsilon_0)^{-1/2}$ is the same as that of c.

(b) Using the definition of the ampere, obtain a value for μ_0 from theoretical considerations. [AEB, '93]

6. (a) In an experiment to determine the speed of light in air, light from a point source is reflected from one face of a sixteen-sided mirror M, travels a distance d to a stationary mirror from which it returns and, after a second reflection at M, forms an image of the source on a screen. When M is rotated at certain speeds, the image is still seen in the same position. Explain how this can occur and show that, if the lowest speed of rotation for which the image remains in the same position is n (in revolutions per second), the speed of light, c, is given by

$$c = 32nd.$$

(b) Using the above arrangement, an image is seen on the screen when the speed of rotation is 900 revolutions per second. The speed of rotation is gradually increased until at 1200 revolutions per second the image is again seen. If $c = 3.00 \times 10^8\,\mathrm{m\,s^{-1}}$, calculate a value for d consistent with these figures. What is the lowest speed of rotation for which an image will be seen on the screen? [J]

7. Describe a terrestrial method of measuring the speed of light. Explain precisely what observations are made, and how the speed is calculated from them.

How can the method you describe be adapted to show qualitatively that the speed of light is less in water than in air? Why would it be difficult to make a precise measurement of the speed of light in water?

What is the evidence that the speed of red and blue light is the same in vacuum, but that red light travels faster than blue light in water? [S]

8. A plane mirror rotating at $35\,\mathrm{rev\,s^{-1}}$ reflects a narrow beam of light to a stationary mirror 200 metres away. The stationary mirror reflects the light normally so that it is again reflected from the rotating mirror. The light now makes an angle of 2.0 minutes of arc with the path it would travel if both mirrors were stationary. Calculate the velocity of light.

Give *two* reasons why it is important that an accurate value of the velocity of light should be known. [J]

3

THE BEGINNINGS OF THE QUANTUM THEORY

3.1 THE BLACK BODY

A black body is a body which absorbs <u>all</u> the radiation which is incident on it.

The concept is an <u>idealized</u> one, but it can be very nearly realized in practice – Fig. 3.1 illustrates how. The inner wall of the enclosure is matt black so that most of any radiation which enters through the small hole is absorbed on reaching the wall. The small amount of radiation which is reflected has very little chance of escaping through the hole before it too is absorbed in a subsequent encounter with the wall.

Fig. 3.1
Approximate realization
of a black body

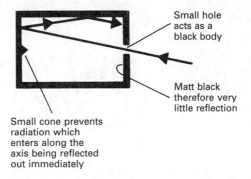

Small hole
acts as a
black body

Matt black
therefore very
little reflection

Small cone prevents
radiation which
enters along the
axis being reflected
out immediately

A black-body radiator (**or cavity radiator**) is one which emits radiation which is <u>characteristic of its temperature</u> and, in particular, which does not depend on the nature of its surfaces.

A black-body radiator can be made by surrounding the enclosure of Fig. 3.1 with a heating coil. The radiation which is emitted by any section of the wall is involved in many reflections before it eventually emerges from the hole. Any section which is a poor emitter absorbs very little of the radiation which is incident on it*, and those sections which are good emitters absorb most of the radiation incident on them. This has the effect of mixing the radiations before they emerge, and of making the temperature the same at all points on the inner surface of the enclosure.

Stars are black-body radiators and therefore their surface temperatures can be estimated by spectroscopic examination of the radiation they emit.

*It is a general result that a good absorber is a good emitter.

3.2 ENERGY DISTRIBUTION IN THE SPECTRUM OF A BLACK BODY

Fig. 3.2 illustrates the way in which the energy radiated by a black body is distributed amongst the various wavelengths. E_λ is such that $E_\lambda \delta\lambda$ represents the energy radiated per unit time per unit surface area of the black body in the wavelength interval λ to $\lambda + \delta\lambda$. It follows that **the area under any particular curve is the total energy radiated per unit time per unit surface area at the corresponding temperature**.

Fig. 3.2
Energy distribution of a black body

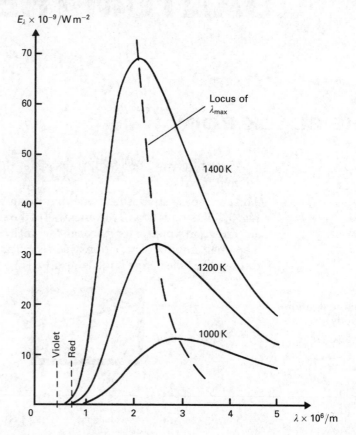

The curves embody two important laws.

Wien's Displacement Law

The wavelength λ_{max} at which the maximum amount of energy is radiated decreases with temperature and is such that

$$\lambda_{max} T = \text{a constant} \tag{3.1}$$

where T is the temperature of the black body in kelvins. Equation [3.1] is known as **Wien's displacement law**. The value of the constant is found by experiment to be $2.90 \times 10^{-3}\,\text{m K}$.

The curves illustrate the well known observation that the colour of a body which is hot enough to be emitting visible light depends on its temperature. At about 1200 K the visible wavelengths which are emitted lie predominantly at the red end of the spectrum and a body at this temperature is said to be red-hot. At higher temperatures the proportions of the other spectral colours increase so that increasing temperatures cause the overall colour to change from red through yellow to white. The intensity distribution of the wavelengths emitted by the Sun is the same as that of a black body at about 6000 K, i.e. the temperature of the Sun's surface is about 6000 K. Some stars are much hotter than the Sun and appear blue.

Stefan's Law

> The total energy radiated per unit time per unit surface area of a black body is proportional to the fourth power of the temperature of the body expressed in kelvins.

Thus

$$E = \sigma T^4 \qquad [3.2]$$

where

σ = a constant of proportionality known as **Stefan's constant**. Its value is $5.67 \times 10^{-8} \, \text{W} \, \text{m}^{-2} \, \text{K}^{-4}$.

Note that the value of E at any temperature T is equal to the area under the corresponding curve, i.e. $E = \int_0^\infty E_\lambda \mathrm{d}\lambda$.

If a black body whose temperature is T is in an enclosure at a temperature T_0, the rate at which unit surface area of the black body is receiving radiation from the enclosure is σT_0^4. The net rate of loss of energy by the black body is therefore given by E_{net} where

$$E_{\text{net}} = \sigma(T^4 - T_0^4) \qquad [3.3]$$

In the case of a non-black body equations [3.2] and [3.3] are replaced by

$$E = \varepsilon \sigma T^4$$

and

$$E_{\text{net}} = \varepsilon \sigma(T^4 - T_0^4)$$

where ε is called the **total emissivity** of the body. Its value depends on the nature of the surface of the body and lies between 0 and 1.

EXAMPLE 3.1

A 100 W electric light bulb has a filament which is 0.60 m long and has a diameter of 8.0×10^{-5} m. Estimate the working temperature of the filament if its total emissivity is 0.70. (Stefan's constant = $5.7 \times 10^{-8} \, \text{W} \, \text{m}^{-2} \, \text{K}^{-4}$.)

Solution

The surface area of the filament is that of a cylinder of diameter 8.0×10^{-5} m and length 0.60 m and is therefore $\pi \times 8.0 \times 10^{-5} \times 0.60 = 1.51 \times 10^{-4} \, \text{m}^2$.

The bulb is rated at 100 W and therefore E, the energy radiated per unit time per unit surface area of the filament, is given by

$$E = \frac{100}{1.51 \times 10^{-4}} = 6.62 \times 10^5 \, \text{W m}^{-2}$$

But

$$E = \varepsilon \sigma T^4$$

$$\therefore \quad 6.62 \times 10^5 = 0.70 \times 5.7 \times 10^{-8} \times T^4$$

i.e. $\quad T^4 = 16.6 \times 10^{12}$

$$\therefore \quad T = 2018 \, \text{K} \approx 2.0 \times 10^3 \, \text{K}$$

3.3 PROCEDURE FOR OBTAINING THE BLACK-BODY ENERGY DISTRIBUTION CURVES

Curves similar to those of Fig. 3.2 (but not extending into the visible region) were obtained by Lummer and Pringsheim in a series of experiments beginning in 1897. In one experiment the black-body radiator was a heated copper sphere (Fig. 3.3) which had been blackened on the inside. A thermocouple was used to measure the temperature of the sphere.

Fig. 3.3
Lummer and Pringsheim's apparatus (schematic) for obtaining black-body radiation curves

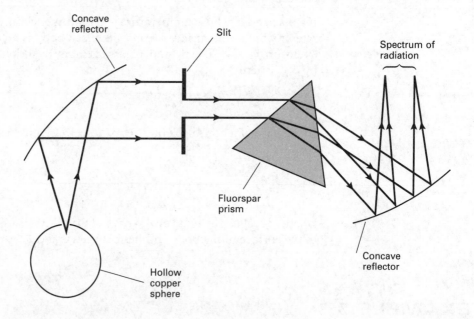

Radiation emerging from a small hole in the wall of the sphere fell onto a concave reflector which directed it, as a <u>parallel</u> beam, towards a fluorspar prism that dispersed the radiation into its spectrum. (Fluorspar is transparent to infrared over the range of wavelengths that were being investigated.)

A **linear bolometer** (Fig. 3.4) was used as a detector. This is a narrow strip of blackened platinum foil. Radiation focused on the strip heated it and increased its electrical resistance. By making the bolometer one arm of a Wheatstone bridge, the energy per unit area per unit time, $E_\lambda \delta \lambda$, of the radiation could be determined over the small range of wavelengths, $\delta \lambda$, intercepted by the bolometer strip. Moving the bolometer to different parts of the spectrum allowed measurements to be made at different wavelengths.

Fig. 3.4
Use of linear bolometer

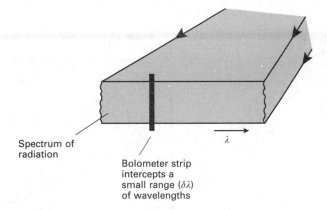

Spectrum of
radiation

Bolometer strip
intercepts a
small range ($\delta\lambda$)
of wavelengths

3.4 THE FAILURE OF CLASSICAL PHYSICS TO ACCOUNT FOR BLACK-BODY RADIATION

Towards the end of the nineteenth century there were several attempts, based on classical physics, to find a formula for the energy distribution of a black body – they all failed. According to classical physics the walls of a black-body radiator can be regarded as containing a number of simple harmonic oscillators that emit

Lord Rayleigh
(1842–1919)

electromagnetic radiation over a continuous range of wavelengths. Further-more, these oscillators were supposed to emit energy in any amount up to some maximum value. Working on these lines, Lord Rayleigh derived a formula which, after a minor modification by Jeans, became known as the **Rayleigh–Jeans formula**

$$E_\lambda = \frac{2\pi c k T}{\lambda^4}$$

[3.4]

where c is the speed of light and k is Boltzmann's constant.

The Rayleigh–Jeans formula agrees quite well with data obtained at long wavelengths, but it fails drastically in other respects.

Fig. 3.5
Comparison between the
Rayleigh–Jeans formula
and experimental data

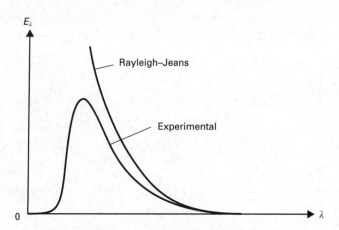

(i)   The experimental curves (Fig. 3.2) fall to zero at short wavelengths, whereas those based on the Rayleigh–Jeans formula head off towards infinity (Fig. 3.5). This became known as the **ultraviolet catastrophe**.

(ii) It is a simple matter to show that on the basis of equation [3.4]

$$\int_0^\infty E_\lambda \mathrm{d}\lambda = \infty$$

This implies that the total amount of energy radiated per unit time per unit surface area over all wavelengths at any single temperature is infinite. This clearly cannot be, and in any case it contradicts Stefan's law.

3.5 PLANCK'S RADIATION LAW

In December 1900 Max Planck came up with a formula that was in complete agreement with experiment at all wavelengths. It is known as **Planck's radiation law**

$$E_\lambda = \frac{2\pi hc^2}{\lambda^5}\left(e^{hc/\lambda kT} - 1\right)^{-1} \qquad\qquad [3.5]$$

where h is Planck's constant ($= 6.626 \times 10^{-34}\,\mathrm{J\,s}$).

Planck had obtained the general form of the law ($E_\lambda = A\lambda^{-5}/(e^{-B/\lambda T} - 1)$ where A and B are constants) in October 1900 and had made several attempts to justify it on the basis of classical physics. They all failed and he was forced to conclude that the laws of classical physics do not necessarily apply on an atomic scale. He proposed instead that **energy can be radiated (or absorbed) only in discrete (separate) packets of energy (now called quanta), each of energy hf where f is the frequency of the radiation concerned**.

This revolutionary proposal marked the beginning of the quantum theory. It implies that **the energy of a system can change only in discrete amounts**, not continuously as had previously been assumed.

Because h is small, the gaps between the allowed energies of any large-scale system are very much smaller than the total energy of the system, and this gives the impression of a continuous range of energies. On the atomic scale, however, the effect is very marked and has major implications.

Max Planck (1858–1947)

Notes (i) **Planck's radiation law approximates to the Rayleigh–Jeans formula at long wavelengths.** This is easily demonstrated.

When λ is large, $hc/\lambda kT \ll 1$ and $e^{hc/\lambda kT}$ can be expanded to give*

$$e^{hc/\lambda kT} \approx 1 + \frac{hc}{\lambda kT}$$

in which case equation [3.5] approximates to

$$E_\lambda = \frac{2\pi hc^2}{\lambda^5} \left(\frac{hc}{\lambda kT} \right)^{-1}$$

i.e. $E_\lambda = \dfrac{2\pi ckT}{\lambda^4}$

which is the Rayleigh–Jeans formula.

(ii) **Planck's radiation law contains Stefan's law and Wien's displacement law**.

The total energy radiated per unit time per unit area, E, is given by

$$E = \int_0^\infty E_\lambda \mathrm{d}\lambda$$

Although it is beyond the scope of this book, it can be shown that substituting Planck's expression (equation [3.5]) for E_λ gives

$$\int_0^\infty E_\lambda \mathrm{d}\lambda = \frac{2\pi^5 k^4}{15c^2 h^3} T^4$$

$$\therefore \quad E = \frac{2\pi^5 k^4}{15c^2 h^3} T^4, \quad \text{i.e.} \quad E \propto T^4$$

*In general

$$e^x = 1 + x + \frac{x}{2!} + \frac{x^3}{3!} + \dots$$

If $|x| \ll 1$, terms in $x^2, x^3, \dots$ may be ignored and therefore

$$e^x \approx 1 + x$$

which is **Stefan's law**. The reader might like to confirm that $2\pi^5 k^4/15c^2h^3 = 5.67 \times 10^{-8}\,\mathrm{W\,m^{-2}\,K^{-4}}$ – the value of σ given in section 3.2.

The wavelength $\lambda_{\max}$ at which the maximum value of E_λ occurs for any temperature T can be found by putting $\mathrm{d}E_\lambda/\mathrm{d}\lambda = 0$. It can be shown* that on the basis of equation [3.5] this leads to

$$\lambda_{\max} = \frac{hc}{4.965\,kT}, \qquad \text{i.e.} \quad \lambda_{\max}T = \text{a constant}$$

which is **Wien's displacement law**. The reader can confirm that $hc/4.965\,k = 2.90 \times 10^{-3}\,\mathrm{m\,K}$ – the value of the constant given in section 3.2.

(iii) Equations [3.4] and [3.5] are often expressed in terms of frequency rather than wavelength. If $E_f\delta f$ represents the energy radiated per unit time per unit area in the frequency range $f \to f + \delta f$, where f is the frequency corresponding to the wavelength λ, then

$$E_f\delta f = -E_\lambda\delta\lambda$$

(The minus sign is included because f decreases when λ increases.) In the limit

$$E_f = -E_\lambda\frac{\mathrm{d}\lambda}{\mathrm{d}f}$$

Since $\lambda = c/f, \qquad \mathrm{d}\lambda/\mathrm{d}f = -c/f^2$

$$\therefore \quad E_f = \frac{c}{f^2}E_\lambda$$

It follows that equations [3.4] and [3.5] can be written as

$$E_f = \frac{2\pi kTf^2}{c^2} \qquad \textbf{(The Rayleigh–Jeans formula)}$$

$$E_f = \frac{2\pi hf^3}{c^2}\,(e^{hf/kT} - 1)^{-1} \qquad \textbf{(Planck's radiation law)}$$

3.6 THE ELECTRONVOLT

The electronvolt (eV) is a unit of energy. It is equal to the kinetic energy gained by an electron in being accelerated by a potential difference of one volt.

The work done when a particle of charge Q moves through a PD V is QV. The charge on the electron is $1.6 \times 10^{-19}\,\mathrm{C}$, and therefore when an electron is accelerated through a PD of $1\,\mathrm{V}$ the work done is

$$(1.6 \times 10^{-19}) \times (1)$$

i.e. $1.6 \times 10^{-19}\,\mathrm{J}$

---

*The method involves solving an equation of the form $x = 5(1 - e^{-x})$. The solution ($x \approx 4.965$) can be found by trial and error.

The work done is equal to the kinetic energy gained by the electron, and therefore the kinetic energy gained by an electron in being accelerated through one volt is 1.6×10^{-19} J,

i.e.

$$1\,eV = 1.6 \times 10^{-19}\,J$$

EXAMPLE 3.2

Write down the kinetic energy, in eV, of (a) an electron accelerated from rest through a PD of 10 V, (b) a proton accelerated from rest through a PD of 20 V, (c) a doubly charged calcium ion accelerated from rest through a PD of 30 V.

Solution

(a) 10 eV   (Since an electron accelerated through 1 V would gain 1 eV, it follows from $W = QV$ that an electron accelerated through 10 V gains 10 eV.)

(b) 20 eV   (Since the charge on the proton has the same magnitude as the charge on the electron, and an electron accelerated through 20 V would gain 20 eV.)

(c) 60 eV   (Since the charge on the calcium ion is twice that on an electron, and an electron accelerated through 30 V would gain 30 eV.)

QUESTIONS 3A

1. An electron is accelerated from rest through a PD of 1000 V. What is (a) its kinetic energy in eV, (b) its kinetic energy in joules, (c) its speed? ($1\,eV = 1.602 \times 10^{-19}$ J, mass of electron $= 9.110 \times 10^{-31}$ kg.)

2. An electron accelerated from rest through a PD of 50 V acquires a speed of $4.2 \times 10^{6}\,m\,s^{-1}$. Without performing a detailed calculation, write down the speed that a PD of 200 V would produce.

3. What is the kinetic energy, in eV, of a triply charged ion of iron (Fe^{3+}) which has been accelerated from rest through a PD of 100 V?

4. An electron moves between a pair of electrodes in a vacuum tube. The first electrode is at a potential of 50 V, and the electron has kinetic energy of 20 eV as it leaves it. What is the potential of the second electrode if the electron just reaches it?

5. The kinetic energy of an α-particle from a radioactive source is 4.0 MeV. What is its speed?
(Charge on electron $= 1.6 \times 10^{-19}$ C, mass of α-particle $= 6.4 \times 10^{-27}$ kg.)

3.7 THE PHOTOELECTRIC EFFECT

Electromagnetic radiation (usually visible light or ultraviolet) incident on a metal surface can cause electrons to be emitted from the surface. The phenomenon is called the **photoelectric effect**. (A broad outline of the experiments which established the nature of the effect is given in section 3.10.)

The conclusions which can be drawn from detailed investigations of the effect are summarized below.

(i)   Emission occurs only if the frequency of the incident radiation is above some minimum value called the **threshold frequency**, and this depends on the particular metal being irradiated. (For example, the threshold frequency of sodium is in the yellow region of the visible spectrum, that of zinc is in the ultraviolet.)

(ii)   Emission commences at the instant the surface starts to be irradiated.

(iii) If the incident radiation is of a single frequency (above the threshold frequency), the number of electrons emitted per second is proportional to the intensity of the radiation.

(iv)   The emitted electrons have various kinetic energies, ranging from zero up to some maximum value. Increasing the frequency of the incident radiation increases the energies of the emitted electrons and, in particular, increases the maximum kinetic energy.

(v)   The intensity of the radiation has no effect on the kinetic energies of the emitted electrons.

3.8 THE INABILITY OF THE WAVE THEORY TO ACCOUNT FOR PHOTOELECTRIC EMISSION

The photoelectric effect is due to electrons absorbing energy from the incident radiation and so becoming able to overcome the attractive forces of the nuclei. According to the wave theory of light the energy of the incident radiation is distributed uniformly over the wavefront. On the basis of the wave theory, therefore, each electron in the surface of an irradiated metal would absorb an equal share of the radiant energy. It is to be expected, therefore, that if the intensity of the radiation were very low, no single electron would gain sufficient energy to escape, or at least, that a considerable time would elapse before any electron did escape. Neither of these predictions is consistent with observation. Furthermore, an increase in intensity increases the energy falling on the surface and would be expected to increase the energies of the emitted electrons. This also is inconsistent with observation. The wave theory can offer no explanation of the frequency dependence of the kinetic energies of the emitted electrons, nor why there should be a minimum frequency at which emission occurs.

3.9 EINSTEIN'S THEORY OF THE PHOTOELECTRIC EFFECT

In 1900 Max Planck had shown that the energy distribution in the black-body spectrum (section 3.5) could be accounted for by assuming that the radiation was emitted as discrete (separate) packets of energy known as **quanta**, the energy, E, of a **quantum** being given by

$$E = hf$$

where

$h = $ a constant, now called **Planck's constant**, equal to $6.626 \times 10^{-34}\,\mathrm{J\,s}$

$f = $ the frequency of the radiation.

In 1905 Einstein extended this idea by suggesting that the quantum of energy emitted by an atom continues to exist as a concentrated packet of energy. A beam of light of frequency f can therefore be considered to be a stream of particles (called **photons**), each of energy hf. At large distances from a point source of light the intensity is low because the photons are spread over a large area, but there is no diminution in the energy associated with each photon. (**Note.** The intensity of a beam of light is proportional to the number of photons per unit cross-section of the beam per unit time.)

Einstein proposed that when a photon collides with an electron, it must either be reflected with no reduction in energy, or it must give up <u>all</u> its energy to the electron. The energy of a single photon cannot be shared amongst the electrons – no more than one electron can absorb the energy of one photon. It follows that (in a given time) **the number of electrons emitted by a surface is proportional to the number of incident photons, i.e. to the intensity of the radiation.** Furthermore, an electron can be emitted as soon as a photon reaches the surface, explaining why photoemission begins instantaneously.

Einstein reasoned that some of the energy imparted by a photon is actually used to <u>release</u> an electron from the surface (i.e. to overcome the binding forces) and that the rest appears as the kinetic energy of the emitted electron. This is summed up by **Einstein's photoelectric equation**

$$hf = W + \tfrac{1}{2}mv^2$$

where

$hf = $ the energy of each incident photon of frequency f,

$W = $ the **work function** of the surface, i.e. the minimum amount of energy that has to be given to an electron to release it from the surface,

$\tfrac{1}{2}mv^2 = $ the **maximum** kinetic energy of the emitted electrons. (Many of the emitted electrons are involved in collisions on their way out of the surface and therefore emerge with energy which is less than the maximum.)

That there should be a minimum frequency which causes emission follows immediately. If $hf < W$, there is not sufficient energy to release an electron, i.e. the threshold frequency, f_0, is given by

$$hf_0 = W$$

The corresponding <u>maximum</u> wavelength, λ_0, is given by

$$h\frac{c}{\lambda_0} = W$$

3.10 MILLIKAN'S VERIFICATION OF EINSTEIN'S PHOTOELECTRIC EQUATION AND MEASUREMENT OF *h*

The photoelectric effect was discovered by Hertz in 1887. He had observed that a spark passes more easily between two electrodes if they are illuminated by ultraviolet radiation. Within a year of Hertz announcing his discovery, Hallwachs found that a negatively charged zinc plate loses its charge when it is illuminated by ultraviolet radiation, but that there is no effect with a positively charged plate. He concluded that the illumination causes the negatively charged plate to emit negatively charged particles. In 1899 Lenard established that these particles were electrons.* By 1902 he had shown that although the maximum kinetic energy of the emitted electrons was independent of the intensity of the incident radiation, it was affected by the frequency being used and by the nature of the illuminated surface. He also showed that the number of electrons emitted was proportional to the intensity of the radiation. It was these experimental results that Einstein explained in 1905.

Lenard's results were not sufficiently accurate to be considered a thorough test of Einstein's theory. Since the theory had such far-reaching consequences, it was of vital importance that it be tested rigorously, and therefore in 1916 Millikan carried out a series of refined experiments – these completely verified Einstein's theory.

A simplified form of the apparatus used by Millikan is shown in Fig. 3.6. He chose to test Einstein's equation by irradiating sodium, potassium and lithium, each of which has a loosely bound electron and therefore exhibits the photoelectric effect over a large range of visible frequencies. Photoelectric emission is a surface phenomenon, and therefore meaningful results can be obtained only if the emitting surfaces are chemically clean. The metals which Millikan used are very reactive and oxidize rapidly in the presence of air, and therefore the apparatus was evacuated. Any oxide that formed on the samples could be removed with the knife (B). The position of the knife was adjusted immediately before a sample was illuminated, so that by rotating the table (T) the surface of the metal could be pared off by moving it against the knife.

Fig. 3.6
Millikan's apparatus to verify Einstein's photoelectric equation

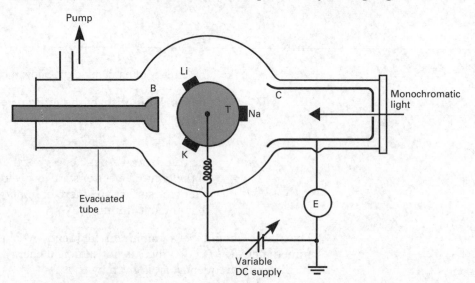

The target metal was irradiated with monochromatic light. The emitted electrons were collected by the electrode (C) and flowed to earth through the electrometer (E) – the electrometer therefore registered a current. (The electrode was coated with copper oxide, a material which shows no photoelectric emission with visible light.)

*The electron had been discovered two years earlier by J. J. Thomson (see section 1.5).

By applying a positive potential to the target metal, Millikan was able to slow down the electrons. This prevented those that had been emitted with low values of kinetic energy from reaching the electrode, and the current through the electrometer decreased. He increased the potential until even those electrons which had been emitted with the maximum kinetic energy were unable to reach the electrode, whereupon the current fell to zero. The minimum potential which reduces the current to zero is called the **stopping potential** V. The work done by an electron in moving against the stopping potential is eV, where e is the charge on the electron. Since this work is done at the expense of the kinetic energy of the electron, it follows that

$$eV = \tfrac{1}{2}mv^2$$

where $\tfrac{1}{2}mv^2 =$ the maximum KE of an emitted electron.

By Einstein's photoelectric equation

$$hf = W + \tfrac{1}{2}mv^2$$

$$\therefore \quad hf = W + eV$$

i.e. $$V = \frac{h}{e}f - \frac{W}{e}$$

Thus, for any given target material, a plot of V against f is linear if Einstein's photoelectric equation is correct. Millikan irradiated each sample with beams of different frequencies and measured the corresponding stopping potentials. For each metal he obtained a graph of the form shown in Fig. 3.7, thus verifying Einstein's equation. The results also provided an accurate value of h.

Fig. 3.7
Graph to verify Einstein's photoelectric equation

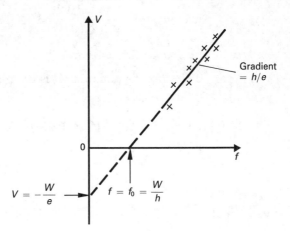

Notes (i) Changing the target metal gives a graph with the same gradient (h/e) but different intercepts.

(ii) The measurements also provide values of W. These are in good agreement with those found from thermionic emission (see Table 1.1).

EXAMPLE 3.3

Sodium has a work function of 2.3 eV. Calculate: (a) its threshold frequency, (b) the maximum velocity of the photoelectrons produced when the sodium is illuminated by light of wavelength 5×10^{-7} m, (c) the stopping potential with light of this wavelength. ($h = 6.6 \times 10^{-34}$ J s, $c = 3.0 \times 10^{8}$ m s^{-1}, 1eV $= 1.6 \times 10^{-19}$ J, mass of electron $m = 9.1 \times 10^{-31}$ kg.)

Solution

(a)   The threshold frequency f_0 is given by

$$hf_0 = W$$

Therefore, since $2.3\,\text{eV} = 2.3 \times 1.6 \times 10^{-19}\,\text{J}$

$$6.6 \times 10^{-34} f_0 = 2.3 \times 1.6 \times 10^{-19}$$

i.e. $f_0 = 5.6 \times 10^{14}\,\text{Hz}$

(b)   By Einstein's photoelectric equation

$$hf = W + \tfrac{1}{2}mv^2$$

$$\therefore \quad h\frac{c}{\lambda} = W + \tfrac{1}{2}mv^2$$

$$\therefore \quad \frac{6.6 \times 10^{-34} \times 3 \times 10^8}{5 \times 10^{-7}} = 2.3 \times 1.6 \times 10^{-19} + \tfrac{1}{2}mv^2$$

$$\therefore \quad 3.96 \times 10^{-19} = 3.68 \times 10^{-19} + \tfrac{1}{2}mv^2$$

i.e. $\tfrac{1}{2}mv^2 = 0.28 \times 10^{-19}$

Therefore the maximum velocity v is given by

$$v = \sqrt{\frac{2 \times 0.28 \times 10^{-19}}{m}}$$

i.e. $v = \sqrt{\dfrac{0.56 \times 10^{-19}}{9.1 \times 10^{-31}}}$

i.e. $v = 2.5 \times 10^5\,\text{m s}^{-1}$

(c)   The stopping potential V is given by

$$eV = \tfrac{1}{2}mv^2$$

i.e. $1.6 \times 10^{-19}\,V = 0.28 \times 10^{-19}$

i.e. $V = 0.18\,\text{V}$

QUESTIONS 3B

1. Calculate the energy of **(a)** a photon of frequency $7.0 \times 10^{14}\,\text{Hz}$, **(b)** a photon of wavelength $3.0 \times 10^{-7}\,\text{m}$.
    ($h = 6.6 \times 10^{-34}\,\text{J s}$, $c = 3.0 \times 10^8\,\text{m s}^{-1}$.)

2. Calcium has a work function of $2.7\,\text{eV}$. **(a)** What is the work function of calcium expressed in joules? **(b)** What is the threshold frequency for calcium? **(c)** What is the maximum wavelength that will cause emission from calcium?
    ($e = 1.6 \times 10^{-19}\,\text{C}$, $h = 6.6 \times 10^{-34}\,\text{J s}$, $c = 3.0 \times 10^8\,\text{m s}^{-1}$.)

3. Gold has a work function of $4.9\,\text{eV}$. **(a)** Calculate the maximum kinetic energy, in joules, of the electrons emitted when gold is illuminated with ultraviolet radiation of frequency $1.7 \times 10^{15}\,\text{Hz}$. **(b)** What is this energy expressed in eV? **(c)** What is the stopping potential for these electrons?
    ($e = 1.6 \times 10^{-19}\,\text{C}$, $h = 6.6 \times 10^{-34}\,\text{J s}$.)

4. Calculate the stopping potential for a platinum surface irradiated with ultraviolet light of wavelength $1.2 \times 10^{-7}\,\text{m}$. The work function of platinum is $6.3\,\text{eV}$.
    ($h = 6.6 \times 10^{-34}\,\text{J s}$, $c = 3.0 \times 10^8\,\text{m s}^{-1}$, $e = 1.6 \times 10^{-19}\,\text{C}$.)

CONSOLIDATION

A black body is a body which absorbs <u>all</u> the radiation incident on it.

The radiation emitted by a black body <u>depends only on its temperature</u>.

Classical physics is unable to account for the energy distribution of a black body.

Planck's Radiation Law

$$E_\lambda = \frac{2\pi hc^2}{\lambda^5} \, (e^{hc/\lambda kT} - 1)^{-1}$$

(i)   Planck's law fits the experimental data at all wavelengths.

(ii) It approximates to the Rayleigh–Jeans formula at long wavelengths.

(iii) It is totally consistent with Stefan's law and Wien's displacement law.

(iv) It cannot be derived on the basis of classical physics. It requires that energy is radiated (and absorbed) in discrete packets of energy E where

$$E = hf$$

and implies that the energy of a system can change only in discrete amounts.

$$E_\lambda = \frac{2\pi ckT}{\lambda^4} \qquad \textbf{(The Rayleigh–Jeans formula)}$$

$$\lambda_{\max} T = \text{a constant} \qquad \textbf{(Wien's displacement law)}$$

$$E = \sigma T^4 \qquad \textbf{(Stefan's law)}$$

The Rayleigh–Jeans formula failed because it approaches infinity at short wavelengths – the **ultraviolet catastrophe,** and predicts that the total energy radiated at any single temperature is infinite.

The electronvolt is a unit of energy equal to the kinetic energy gained by an electron in being accelerated through a PD of 1 volt.

In order to explain the photoelectric effect a beam of light (of frequency f and wavelength λ) is regarded as a stream of **photons**, each of energy E where

$$E = hf \qquad \text{or} \qquad E = h\frac{c}{\lambda}$$

The Photoelectric Effect

Electromagnetic radiation incident on a metal surface causes electrons to be emitted from the surface.

The maximum KE of the emitted electrons increases when the **frequency** of the radiation increases.

The number of emitted electrons is proportional to the **intensity** of the radiation.

The work function of a surface is the minimum amount of energy that has to be given to an electron to release it from the surface.

The threshold frequency is the minimum frequency that will cause emission.

Einstein's Photoelectric Equation

$$hf = W + \tfrac{1}{2}mv^2$$

Since hf can be replaced by hc/λ, W by hf_0, and $\tfrac{1}{2}mv^2$ by eV, where $V =$ stopping potential and $f_0 =$ threshold frequency

$$\begin{pmatrix} hf \\ \text{or} \\ h\dfrac{c}{\lambda} \end{pmatrix} = \begin{pmatrix} W \\ \text{or} \\ hf_0 \end{pmatrix} + \begin{pmatrix} \tfrac{1}{2}mv^2 \\ \text{or} \\ eV \end{pmatrix}$$

Millikan's Experiment

(i)   Verified Einstein's photoelectric equation.

(ii)   Provided an accurate value of h.

(iii)   Provided values of the work functions of the metals used.

$$hf = W + \tfrac{1}{2}mv^2$$

$$\therefore \quad hf = W + eV \qquad \text{where } V = \text{ stopping potential}$$

$$\therefore \quad V = \frac{h}{e}f - \frac{W}{e}$$

The graph of V against f is a straight line (hence (i)) of gradient h/e (hence (ii)) and y-intercept $-W/e$ (hence (iii)).

Changing the target material changes the intercept, but not the gradient.

QUESTIONS ON CHAPTER 3

1. Sketch graphs showing the distribution of energy in the spectrum of black-body radiation at three temperatures, indicating which curve corresponds to the highest temperature. If such a set of graphs were obtained experimentally, how would you use information from them to attempt to illustrate Stefan's Law?   [L]

2. Draw a graph showing the distribution of energy in the spectrum of a black body. Explain what quantity is plotted against the wavelength.
By considering how this energy distribution varies with temperature explain the colour changes which occur when a piece of iron is heated from cold to near its melting point.   [L]

3. The silica cylinder of a radiant wall heater is 0.6 m long and has a radius of 5 mm. If it is rated at 1.5 kW estimate its temperature when operating. State *two* assumptions you have made in making your estimate. (Stefan's constant, $\sigma = 6 \times 10^{-8}\,\text{W}\,\text{m}^{-2}\,\text{K}^{-4}$.)   [L]

4. (a) Explain what is meant by black-body radiation.

(b) Outline a method of measuring the energy distribution in the spectrum of a black-body radiator at a single temperature.

5. (a) The temperature of a piece of wire is gradually increased. Discuss the variation in character of the radiation emitted. Sketch graphs to illustrate this variation. (Assume that the wire behaves like a black body.) Suggest how you might investigate experimentally the variation in the *total* radiation emitted by the wire at the various temperatures. In what way would this total radiation be related to the graphs already sketched?

(b) If the mean equilibrium temperature of the Earth's surface is T and the total rate of energy emission by the Sun is E show that

$$T^4 = \frac{E}{16\,\sigma\,\pi\,R^2}$$

where σ is the Stefan constant and R is the radius of the Earth's orbit around the Sun. (Assume that the Earth behaves like a black body.)   [L]

6. The diagram shows how E_λ, the energy radiated per unit area per second per unit wavelength interval, varies with wavelength λ for radiation from the Sun's surface.

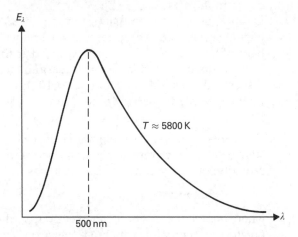

Calculate the wavelengths λ_{max} at which the corresponding curves peak for:
(a) radiation in the Sun's core where the temperature is approximately 15×10^6 K, and
(b) radiation in interstellar space which corresponds to a temperature of approximately 2.7 K
(You may use the relation $\lambda_{max} \times T =$ constant.)
Name the part of the electromagnetic spectrum to which the calculated wavelength belongs in each case. [L]

7. What radical proposal did Max Planck make in order to account for the characteristics of blackbody radiation?

8. The Rayleigh–Jeans formula, which is based on classical physics, fails in two ways to account for the energy distribution of a black body. What are they?

9. The solar radiation falling normally on the surface of the Earth has an intensity 1.40 kW m^{-2}. If this radiation fell normally on one side of a thin, freely suspended blackened metal plate and the temperature of the surroundings was 300 K, calculate the equilibrium temperature of the plate. Assume that all heat interchange is by radiation.

(Stefan's constant $= 5.67 \times 10^{-8} \text{ W m}^{-2} \text{K}^{-4}$.) [L]

10. The total power output of the Sun is 3.79×10^{26} W. Calculate

(a) the temperature of the Sun's surface,
(b) the wavelength at which the Sun radiates the maximum energy.
(Diameter of Sun $= 1.39 \times 10^9$ m, $\sigma = 5.67 \times 10^{-8} \text{ W m}^{-2} \text{K}^{-4}$, $\lambda_{max} T = 2.90 \times 10^{-3} \text{ m K}$.)

11. A solid at high temperature emits total (blackbody) radiation. Describe, with the aid of a sketch graph, how the radiated energy is distributed across the emitted wavelength range.
How would the graph change if the absolute temperature of the solid were **(a)** doubled, **(b)** halved? [L, '93]

12.

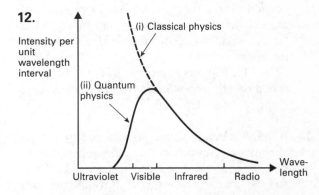

The curves above relate to the radiation from a furnace at a high temperature according (i) to classical physics, and (ii) to quantum physics. The y-axis is the energy per unit wavelength interval; the x-axis is the wavelength.
Which kind of radiation would predominate from this furnace according to **(a)** classical physics, **(b)** quantum physics?
Which feature of curve (ii) was explained by Planck's introduction of the *quantum*?
State *one* other physical quantity which is quantized and identify the corresponding quantum. [L, '94]

13. Describe the principal experimental facts concerning the photoelectric effect, and show how they are explained by the quantum theory.
Describe how Planck's constant h can be determined by experiments on the photoelectric effect. [S]

14. Light of frequency 6.0×10^{14} Hz incident on a metal surface ejects photoelectrons having a kinetic energy 2.0×10^{-19} J.
Calculate the energy needed to remove an electron from the metal (work function).

Very briefly indicate how you would determine experimentally the kinetic energy of the photo-electrons. (The Planck constant $= 6.6 \times 10^{-34}$ J s.)   [S]

15. Light of wavelength $0.50\ \mu m$ incident on a metal surface ejects electrons with kinetic energies up to a maximum value of 2.0×10^{-19} J. What is the energy required to remove an electron from the metal? If a beam of light causes no electrons to be emitted, however great its intensity, what condition must be satisfied by its wavelength? (The Planck constant $= 6.6 \times 10^{-34}$ J s, the speed of light $= 3.0 \times 10^8$ m s^{-1}.)   [S]

16. The maximum kinetic energy of photoelectrons ejected from a tungsten surface by monochromatic light of wavelength 248 nm was found to be 8.6×10^{-20} J. Find the work function of tungsten.
(The Planck constant, $h = 6.6 \times 10^{-34}$ J s; speed of light, $c = 3.0 \times 10^8$ m s^{-1}; electronic charge, $e = -1.6 \times 10^{-19}$ C.)   [C(O)]

17. When a metallic surface is exposed to monochromatic electromagnetic radiation electrons may be emitted. Apparatus is arranged so that **(a)** the intensity (energy per unit time per unit area) and **(b)** the frequency of the radiation may be varied. If each of these is varied in turn whilst the other is kept constant what is the effect on **(i)** the number of electrons emitted per second, and **(ii)** their maximum speed? Explain how these results give support to the quantum theory of electromagnetic radiation.
The photoelectric work function of potassium is 2.0 eV. What potential difference would have to be applied between a potassium surface and the collecting electrode in order just to prevent the collection of electrons when the surface is illuminated with radiation of wavelength 350 nm? What would be **(iii)** the kinetic energy, and **(iv)** the speed, of the most energetic electrons emitted in this case?
(Speed of electromagnetic radiation *in vacuo* $= 3.0 \times 10^8$ m s^{-1}. The electronic charge $= -1.6 \times 10^{-19}$ C. Mass of an electron $= 9.1 \times 10^{-31}$ kg. The Planck constant $= 6.6 \times 10^{-34}$ J s.)   [L]

18. List the important experimental facts relating to the photoelectric effect, and explain how Einstein's equation accounts for them.
A clean surface of potassium in a vacuum is irradiated with light of wavelength $5.5 \times$ 10^{-7} m and electrons are found just to emerge, but when light of wavelength 5×10^{-7} m is incident, electrons emerge each with energy 3.62×10^{-20} J. Estimate the value for Planck's constant h.
Deduce the effect of irradiating in vacuum **(a)** a copper surface, and **(b)** a caesium surface, with light of wavelength 5×10^{-7} m, given that the work functions of copper and caesium are, respectively, 6.4×10^{-19} J and 3.2×10^{-19} J.
(Velocity of light $= 3 \times 10^8$ m s^{-1}.)   [W]

19. Einstein's equation for the photoelectric emission of electrons from a metal surface under radiation of frequency v can be written as

$$hv = \tfrac{1}{2} m_e v^2 + \phi,$$

where m_e is the mass of an electron, v the greatest speed with which an electron can emerge and ϕ is a quantity called the work function of the metal.
(a) Explain briefly the physical process with which this equation is concerned.
(b) Describe briefly an experiment by which you could determine the values of h/e and ϕ.
(c) For sodium the value of ϕ is 3.12×10^{-19} J, and the wavelength of sodium yellow light is 590 nm.
  (i) Explain why electrons are emitted when a sodium surface is irradiated with sodium yellow light, and calculate the greatest speed of the emitted electrons.
  (ii) Estimate the 'stopping potential' for these electrons, assuming that no contact potential differences are involved.
(Take the value of the speed of light in vacuum, c, to be 3.00×10^8 m s^{-1}, the Planck constant, h, to be 6.63×10^{-34} J s, the electronic charge, e, to be -1.60×10^{-19} C, and m_e to be 9.11×10^{-31} kg.)   [O]

20. (a) When electromagnetic radiation falls on a metal surface, electrons may be emitted. This is the photoelectric effect.
  (i) State Einstein's photoelectric equation, explaining the meaning of each term.
  (ii) Explain why, for a particular metal, electrons are emitted only when the frequency of the incident radiation is greater than a certain value.

    (iii) Explain why the maximum speed of the emitted electrons is independent of the intensity of the incident radiation.

(b) A source emits monochromatic light of frequency 5.5×10^{14} Hz at a rate of 0.10 W. Of the photons given out, 0.15% fall on the cathode of a photocell which gives a current of $6.0\,\mu$A in an external circuit. You may assume this current consists of all the photoelectrons emitted. Calculate:

    (i) the energy of a photon,

    (ii) the number of photons leaving the source per second,

    (iii) the percentage of the photons falling on the cathode which produce photo-electrons.

(c) Calculate the wavelength associated with electrons which have been accelerated from rest through 3000 V.

(The Planck constant $= 6.6 \times 10^{-34}$ J s, the electron charge $= 1.6 \times 10^{-19}$ C, the electron mass $= 9.1 \times 10^{-31}$ kg.)   [J, '91]

---

21. (a) Write down an expression for the energy of a photon, explaining the meanings of the symbols used in your expression and giving the units of the physical quantities involved.

(b) In an experiment with a vacuum photocell the maximum kinetic energy of the electrons emitted was measured for different wavelengths of the illuminating radiation. The following results were obtained.

Maximum kinetic energy/10^{-19} J	Wavelength/ 10^{-7} m
3.26	3.00
2.56	3.33
1.92	3.75
1.25	4.29
0.58	5.00

Use these results to plot a linear graph and derive a value for Planck's constant.

(c) If the experiment were repeated with radiation of wavelength **(i)** 7.5×10^{-7} m, **(ii)** 2.8×10^{-7} m, would photoelectrons be emitted and, if so, what would be their maximum kinetic energy?

(d) Describe and explain how the graph might change if a different metal were used for the surface of the photo-cathode.

Speed of light, $c = 3.00 \times 10^8$ m s^{-1}. [J]

---

22. Light of photon energy 3.5 eV is incident on a plane photocathode of work function 2.5 V. Parallel and close to the cathode is a plane collecting electrode. The cathode and collector are mounted in an evacuated tube.

(a) Find the maximum kinetic energy E_{max} of photoelectrons emitted from the cathode. (Express your answer in eV.)

(b) Find the minimum value of the potential difference which should be applied between collector and cathode in order to prevent electrons of energy E_{max} from reaching the collector for electrons emitted

    (i) normal to the cathode,

    (ii) at an angle of 60° to the cathode.

[C]

---

23. (a) (i) Explain what is meant by *photoelectric emission*.

    (ii) Briefly describe a simple experiment to demonstrate this effect qualitatively.

(b) Write down Einstein's photoelectric equation and explain the meaning of each term in it.

(c) In an experiment in photoelectricity, the maximum kinetic energy of the photo-electrons was determined for different wavelengths of the incident radiation. The following results were obtained:

Wavelength/nm	300	375	500
Maximum kinetic energy/eV	2.03	1.20	0.36

Use the results to determine
(i) the work function for the metal,
(ii) a value for Planck's constant.
($c = 3 \times 10^8$ m s^{-1}, $e = 1.6 \times 10^{-19}$ C.)

[W, '90]

---

24. (a) Electrons may be emitted from a surface by *thermionic emission* or by *photoelectric emission*. Distinguish between the two.

(b) (i) Write down Einstein's photoelectric equation relating the maximum kinetic energy E_{max} of the photo-electrons with the frequency f of the incident radiation and with the work function ϕ for the emitting surface.

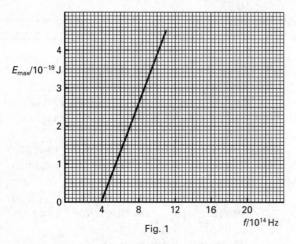

Fig. 1

(ii) The graph shows how E_{max} varies with f for a particular surface. Use the graph to find the value of
(I) ϕ, (II) the threshold wavelength of the radiation.

(iii) Describe and explain the effect of increasing the intensity of the incident radiation.

(c) A photocell of work function $\phi = 1.0\,eV$ is connected to the circuit shown in Fig. 2. The wavelength of radiation incident on the cell is gradually decreased and the output of the amplifier continuously monitored. Fig. 3 gives the variation of wavelength and amplifier output with time. Using Figures 2 and 3 calculate the wavelength at time T.

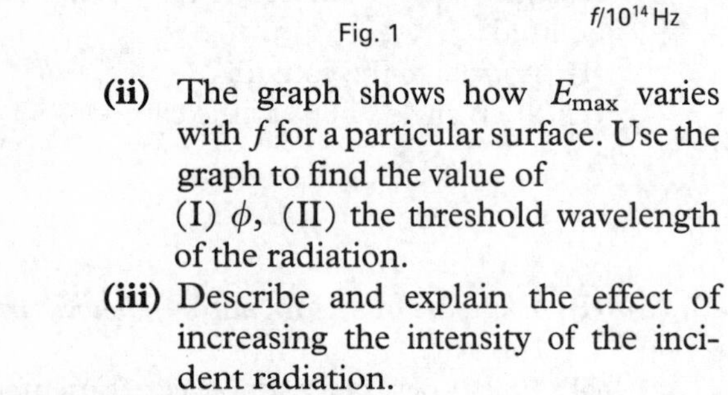

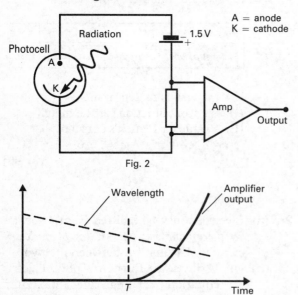

Fig. 2

Fig. 3

(Charge on an electron $e = -1.60 \times 10^{-19}\,C$, Speed of light $c = 3.00 \times 10^8\,m\,s^{-1}$, the Planck constant $h = 6.63 \times 10^{-34}\,J\,s$.)   [W, '92]

25. Explain the physical processes described by the Einstein photoelectric equation $h\nu = \frac{1}{2}mv^2 + h\nu_0$, and state the significance of each term.

Describe briefly how the values of h and ν_0 can be determined.

An argon laser emits a beam of light of wavelength $4.88 \times 10^{-7}\,m$, the power in the beam being 100 mW. How many photons per second are emitted by the laser? If the beam falls on the caesium cathode of a photocell, what photoelectric current would be observed, assuming 10% of the photons are able to eject an electron? Given that the limiting frequency ν_0 of caesium is $5.2 \times 10^{14}\,Hz$, what reverse potential difference between the cell electrodes is needed to suppress the photocell current?

(The Planck constant $= 6.6 \times 10^{-34}\,J\,s$, the speed of light $= 3.0 \times 10^8\,m\,s^{-1}$, the electronic charge $= 1.6 \times 10^{-19}\,C$.)   [S]

26. What is meant by *photoelectricity*? Describe the main features of photoelectric emission. Give an expression for the kinetic energy of the photoelectrons emitted from a surface, explaining what is meant by *work function*, the *threshold frequency* (or *cut-off frequency*) and *Planck's constant*.

Describe an experiment to verify the equation for the kinetic energy of the photoelectrons, and show how the work function of the surface and Planck's constant can be obtained.

A monochromatic light source provides a 5 W beam of radiation of wavelength $4.5 \times 10^{-7}\,m$ and this beam liberates 10^{10} photoelectrons per second from the surface of a sodium block. The threshold wavelength of sodium is $5.5 \times 10^{-7}\,m$.

(a) Calculate the magnitude of the photoelectric emission current (in amperes) given by **(i)** this arrangement, **(ii)** an otherwise identical one with a 10 W beam, and **(iii)** a 5 W beam of wavelength $6 \times 10^{-7}\,m$.

(b) If, in the original arrangement, the sodium block were electrically isolated, the emission of photoelectrons would cause it to acquire a positive potential. Find the steady value which it would eventually reach.

(Electronic charge $= 1.6 \times 10^{-19}\,C$; the Planck constant $= 6.6 \times 10^{-34}\,J\,s$; speed of light $= 3 \times 10^8\,m\,s^{-1}$.)   [W]

27. (a) **(i)** What is a *photon*?

(ii) Show that E, the energy of a photon, is related to λ, its wavelength, by

$$E\lambda = 1.99 \times 10^{-16}$$

where E is measured in J and λ is measured in nm.

(b) Two metal electrodes A and B are sealed into an evacuated glass envelope and a potential difference V, measured using the voltmeter, is applied between them as shown in Fig. 1.

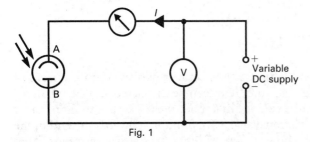

Fig. 1

B is then illuminated with monochromatic light of wavelength 365 nm and I, the current in the circuit, is measured for various values of V. The results are shown in Fig. 2.

(i) From this graph, deduce the PD required to stop photoelectric emission from B.

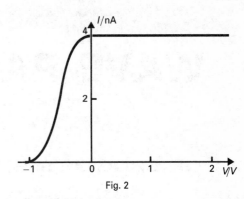

Fig. 2

(ii) Calculate the maximum kinetic energy of the photoelectrons.

(iii) Deduce the work function energy of B.

(The Planck constant $= 6.6 \times 10^{-34}$ J s, the electron charge $= 1.6 \times 10^{-19}$ C, velocity of light $= 3.0 \times 10^{8}$ m s^{-1}.)

[C, '91]

4

WAVE–PARTICLE DUALITY

4.1 DE BROGLIE'S HYPOTHESIS

The reflection and refraction of light are satisfactorily explained on the basis of light being a wave motion. Furthermore, light can be diffracted and can produce interference effects – convincing evidence that it behaves as a wave motion. Light can be polarized; it is therefore a <u>transverse</u> wave motion. Towards the end of the nineteenth century Maxwell showed, on entirely theoretical grounds, that electromagnetic waves could propagate through space; the velocity of these waves is exactly the same as that of light.

It seems paradoxical therefore that it is necessary to attribute particle properties to light in order to explain the photoelectric effect. What the reader must accept, though, is that the idea of light being a wave motion and the idea of it being a particle motion are merely two different models which help us explain the behaviour of light; neither is necessarily a literal description of what light is.

Louis de Broglie
(1892–1987)

In a thesis presented in 1924 Louis de Broglie, having reflected on the wave–particle duality of light, suggested that matter might also have a dual nature. He proposed that any particle of momentum p has an associated wavelength λ (now called the **de Broglie wavelength**) given by

$$\lambda = \frac{h}{p} = \frac{h}{mv}$$

[4.1]

where m is the relativistic mass* of the particle and v is its velocity. This relationship was confirmed in 1927 when Davisson and Germer succeeded in diffracting electrons (see section 4.3).

Neutrons, protons, hydrogen atoms and helium atoms have also been diffracted. Neutron diffraction is used to study crystal structures. The wavelengths associated with macroscopic bodies are very much less than the width of any aperture through which such a body might pass, and any diffraction which occurs is too small to be observable. (The de Broglie wavelength of a snooker ball moving at $1\,\mathrm{m\,s}^{-1}$ is of the order of $10^{-33}\,\mathrm{m}$.)

Note   The kinetic energy, $\frac{1}{2}mv^2$, of an electron which has been accelerated from rest through a potential difference, V, is given by

$$\frac{1}{2}mv^2 = eV$$

$$\therefore \qquad m^2v^2 = 2meV$$

i.e. $$mv = \sqrt{2meV}$$

Since $\lambda = h/mv$

$$\lambda = \frac{h}{\sqrt{2meV}}$$ [4.2]

QUESTIONS 4A

1. Calculate the de Broglie wavelength of an electron moving at $3.0 \times 10^6\,\mathrm{m\,s}^{-1}$.
   ($h = 6.6 \times 10^{-34}\,\mathrm{J\,s}$, mass of electron = $9.1 \times 10^{-31}\,\mathrm{kg}$.)

2. Calculate **(a)** the speed, **(b)** the de Broglie wavelength of an electron which has been accelerated from rest through a PD of 250 V.

   ($e = 1.6 \times 10^{-19}\,\mathrm{C}$, $h = 6.6 \times 10^{-34}\,\mathrm{J\,s}$, mass of electron = $9.1 \times 10^{-31}\,\mathrm{kg}$.)

3. Calculate the de Broglie wavelength of an α-particle of energy 4.0 MeV.
   ($e = 1.6 \times 10^{-19}\,\mathrm{C}$, $h = 6.6 \times 10^{-34}\,\mathrm{J\,s}$, mass of α-particle = $6.4 \times 10^{-27}\,\mathrm{kg}$.)

4.2 INTERPRETATION OF WAVE AMPLITUDE

The de Broglie waves, though often referred to as **matter waves**, are not composed of matter. The amplitude of the wave at any point (x, y, z, t) in space and time is represented by a mathematical function called the **wave function, ψ, of the wave, where ψ is some function of** x, y, z and t, i.e. $\psi = \psi(x, y, z, t)$.

---

*The special theory of relativity distinguishes between the mass m_0 of a stationary particle (the rest mass) and the mass m of a particle moving with velocity v. It can be shown that $m = m_0(1 - v^2/c^2)^{-1/2}$, where c is the velocity of light. The distinction is unimportant if $v \ll c$.

The square of the amplitude of the wave at any given point (in space and time) is equal to the probability of the associated particle being at that point. (Strictly, since ψ may be complex, the probability is equal to $\psi\,\psi^*$, where ψ^* is the complex conjugate of ψ.)

The situation is equivalent to that for light. The probability of a photon being at some point is proportional to the intensity of the light wave at that point, and this is proportional to the square of the amplitude of the electric field vector.

We can no longer think of a particle as being a <u>point</u> because, in general, its wave function will be non-zero over some extended region of space rather than just a single point.

4.3 ELECTRON DIFFRACTION

Confirmation of de Broglie's hypothesis was first provided in 1927 by Davisson and Germer. They showed that the scattering of electrons by a single crystal of nickel could be satisfactorily explained on the basis of the electrons having been diffracted. The regularly spaced atoms in a crystal act as diffraction centres in much the same way as they do for X-rays. (The de Broglie wavelengths of electrons which have been accelerated by potentials in the range 1.5 V to 15 000 V are in the range 10^{-9} m to 10^{-11} m, similar to the wavelengths of X-rays.) Further confirmation was provided by G.P. Thomson in 1928 using a different technique.

The Experiments of Davisson and Germer (1927)

A simplified version of the experimental arrangement used by Davisson and Germer is shown in Fig. 4.1. A focused beam of electrons was directed towards the surface of a nickel crystal in an evacuated enclosure. The electrons were scattered in different directions and could be detected by means of a Faraday cylinder connected to a galvanometer.

Fig. 4.1
Experimental arrangement used by Davisson and Germer to demonstrate electron diffraction

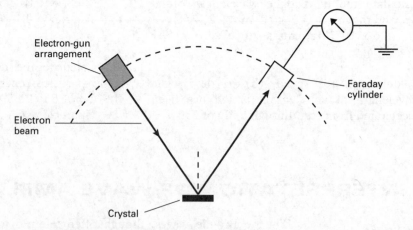

In one version of the experiment the detector was positioned so as to collect only those electrons which had been scattered <u>at the same angle</u> to the normal as the incident beam. (This procedure was originally used by Bragg in experiments on X-ray diffraction and is known as **Bragg reflection**.) The velocity of the incident electrons was gradually increased (by increasing the anode voltage (V) on the electron gun) and the detector reading was found to go through a series of maxima (Fig. 4.2). The results were entirely consistent with the crystal planes having acted

Fig. 4.2
Intensity of scattered
beam as a function of
(accelerating PD)$^{1/2}$

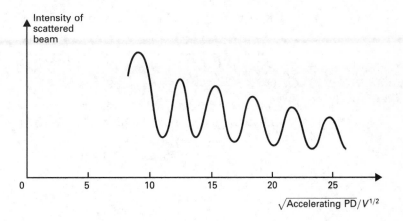

as a reflection grating to diffract the electrons – the various maxima representing different orders of diffraction. Furthermore the wavelengths were found to be in complete agreement with the values predicted on the basis of de Broglie's equation.

Fig. 4.3
(a) To show that waves
scattered by atoms in any
one plane are in phase
with each other,
(b) To establish the
condition that parallel
planes scatter in phase
with each other

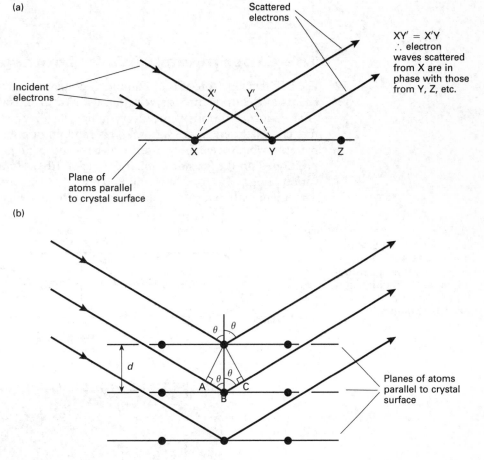

Theory

Because the incident and 'reflected' beams are at the same angle to the normal, the waves from all the atoms in any one plane are bound to be in phase with each other (see Fig. 4.3(a)). Furthermore, the waves from any plane will be in phase with those from every other plane, and therefore produce diffraction maxima, when (in Fig. 4.3(b))

$$AB + BC = n\lambda \qquad (n = 1, 2, 3, \ldots)$$

i.e. $2d \cos \theta = n\lambda$

where λ is the de Broglie wavelength of the electrons and d is the spacing between adjacent planes of atoms. Since

$$\lambda = \frac{h}{\sqrt{2meV}} \qquad \text{(equation [4.2])}$$

$$2d \cos \theta = \frac{nh}{\sqrt{2meV}}$$

$$\therefore \qquad V^{1/2} = \frac{nh}{2d \cos \theta \sqrt{2me}}$$

Since θ is fixed

$$V^{1/2} \propto n$$

which explains why the maxima in Fig. 4.2 are equally spaced.

The Experiments of G. P. Thomson (1928)

G. P. Thomson (son of J. J. Thomson) obtained diffraction patterns by passing beams of electrons through very thin ($\sim 10^{-8}$ m) metal foils in a discharge tube (Fig. 4.4(a)). The foils were polycrystalline, i.e. they consisted of large numbers of tiny crystals oriented at random. This random orientation produced diffraction patterns which were in the form of concentric rings on a photographic plate positioned on the far side of the foil (Fig. 4.4(b)). By measuring the radii of the various rings Thomson was able to show that the wavelength was in excellent agreement with that calculated on the basis of de Broglie's equation.

Fig. 4.4
(a) Experimental arrangement (schematic) used by G.P. Thomson to demonstrate electron diffraction
(b) Electron diffraction pattern of gold foil

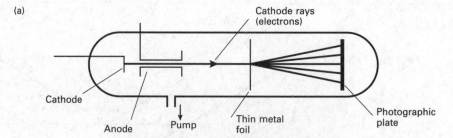

(a)

(b)

4.4 DOUBLE-SLIT INTERFERENCE WITH ELECTRONS

Suppose we were to conduct a Young's double-slit experiment using electrons rather than light (Fig. 4.5).* On examining the photographic film we would find an interference pattern remarkably similar to that produced by a monochromatic source of light, but in which the 'fringes' were of the order of a thousand times closer together.

Fig. 4.5
Thought experiment for double-slit interference with electrons

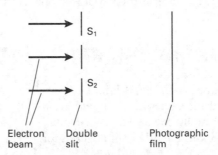

The reader may be tempted to think that half the electrons pass through S_1 and half through S_2, and that it is interference between the two sets of electrons that produces the pattern. This is _not_ what happens. Imagine that the intensity of the electron beam is so low that electrons arrive at the double slit one by one. Providing we leave the photographic film in place long enough, we would still obtain the interference pattern. Clearly, **each electron must pass through both slits and must interfere with itself**. We cannot predict where any one electron will go after it has passed through the slits. The best we can do is calculate the probability of an electron arriving at any particular point on the film. The highest probabilities are associated with the centres of bright fringes; the lowest with the centres of dark fringes, etc. Fig. 4.6 shows what the film might look like at various stages as the pattern is building up.

Fig. 4.6
Build-up of double-slit interference pattern after (a) 28, (b) 1000, (c) 10 000 electrons have contributed to it

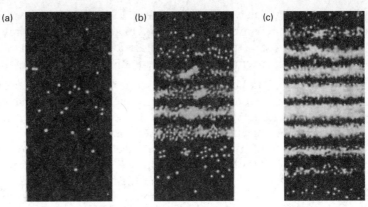

We have stated that each electron passes through both slits. There is clearly a problem with this as long as we regard an electron as a particle and take that to mean that it is a point. If we regard an electron as a wave, there is no problem. We are used to thinking of light as a wave motion and have no difficulty in accepting that it passes through both slits at once in the optical version of the double-slit experiment. What we must accept is that each photon passes through both slits and interferes with itself.

*Although it is not possible to use slits, it is possible to conduct experiments in which electrons behave as if they have passed through a double slit and this allows us to say what would happen if we could carry out the 'thought experiment' we are describing.

The reader who is familiar with the details of Young's experiment will be aware of the need for the light from the two slits to be coherent (i.e. to have a constant phase difference). A little thought should convince him/her that there would be no coherence if it were not for the fact that each photon interferes with itself, because individual photons arrive at the slits at random. The electrons also reach the slits at random and this line of argument therefore supports our earlier assertion that it would be wrong to interpret the pattern as being due to electrons from one slit interfering with electrons from the other slit.

QUESTIONS 4B

1. Suppose that a double-slit experiment is conducted with electrons using the arrangement of Fig. 4.5 but with one slit covered for the first half of the experiment and the other slit covered for the second half. Would the pattern on the photographic film be the same as that which would be obtained if both slits were open for the whole of the experiment? Explain your answer.

4.5 THE TRANSMISSION ELECTRON MICROSCOPE (TEM)

The operation of an electron microscope relies on the fact that beams of electrons can be focused by means of suitably shaped magnetic fields. The fields are produced by electromagnets and are the 'lenses' of the microscope.

Thermionically emitted electrons are accelerated through a PD of 50–100 kV and enter the **condenser lens** which directs them (as a parallel beam) through the sample being studied (Fig. 4.7). The **objective lens** then forms a magnified real image which acts as an object for the **projector lens**. The projector lens creates an even larger image on a fluorescent screen or a photographic plate.

The whole arrangement, including the object, is contained in a highly evacuated chamber. If it were not, electrons would collide with air molecules and this would spoil the image. An air-lock arrangement allows the operator to change the specimen under investigation without losing the vacuum.

The ability of a microscope to resolve fine detail is limited by diffraction effects and therefore depends on wavelength – the smaller the wavelength, the greater this ability, i.e. the greater the **resolving power**. An optical microscope cannot resolve objects which are closer than 2×10^{-7} m. For an electron microscope operating at 100 kV the de Broglie wavelength of the electrons is about 10^5 times smaller than that of (blue) light and we might expect, therefore, that an electron microscope could resolve objects which are as close as 2×10^{-12} m. In practice, the best that can be achieved is about 2×10^{-10} m. There are three main reasons for this.

(i)   The pole pieces of the electromagnets cannot be manufactured to such an accuracy that the field pattern is perfect. This creates an effect analogous to the spherical aberration produced by an optical lens.

(ii)   The electrons are emitted from the cathode with a range of energies and therefore do not all have exactly the same speed. Since the trajectory of an electron depends on its speed, electrons from the same point on the object can end up at different points on the image – an effect analogous to chromatic aberration. **The specimen must be very thin** (typically 50 nm). If it were not, the energy lost by the electrons in passing through it would result in a decrease in resolution because there would be a significant increase in the spread in energy of the emerging electrons.

Fig. 4.7
The transmission
electron microscope
(TEM)

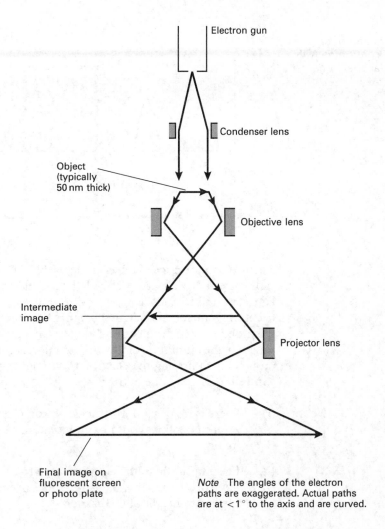

Electron gun

Condenser lens

Object
(typically
50 nm thick)

Objective lens

Intermediate
image

Projector lens

Final image on
fluorescent screen
or photo plate

Note The angles of the electron
paths are exaggerated. Actual paths
are at < 1° to the axis and are curved.

(iii)   The focal length of a magnetic lens depends on the current in its coil. Slight
fluctuations in current are unavoidable and, in effect, decrease the
resolution by adversely affecting the focusing.

There is no point in the magnifying power of a microscope being any greater than
that which just allows the eye to resolve the detail that has been resolved by the
microscope itself. The maximum useful magnifying power of the best optical
microscopes is about $2000\times$. Because of their superior ability to resolve fine detail,
the corresponding figure for a transmission electron microscope is about $10^6\times$.

Note   The paths of the electrons are determined by their <u>particle</u> properties; the <u>wave</u>
nature of the electrons is important only in terms of resolution.

4.6 TUNNELLING

Consider a situation in which the potential energy of an electron varies with
position (x) as shown in Fig. 4.8.

Classically, an electron with energy E which is initially to the left of x_1 could never
move to a point to the right of x_1 because it does not have sufficient energy to do so.
The region between x_1 and x_2 acts as a **potential energy barrier** that the

Fig. 4.8
Wave function of an
electron confronted with
a potential energy barrier

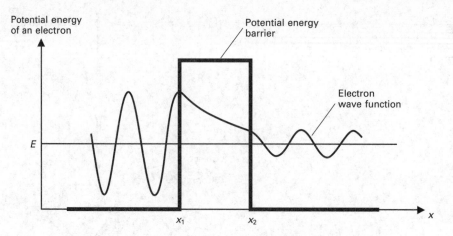

electron cannot cross. However, according to quantum mechanics the electron can be treated as a wave. To the left of x_1 its wave function is sinusoidal and is of large amplitude. Between x_1 and x_2 the wave function decays exponentially, becoming sinusoidal again to the right of x_2 but of much smaller amplitude. Since the probability of the electron being at any particular point is proportional to the square of the amplitude of its wave function at that point, there is a small but finite probability of finding the electron to the right of x_2. The electron is said to have **tunnelled** through the barrier.

Notes (i) There is no possibility of the electron ever being <u>observed</u> between x_1 and x_2 because this would be in violation of the principle of conservation of energy.

(ii) The amplitude of the electron wave to the right of x_2 depends on both the width of the barrier and the difference in energy between the electron and the maximum height of the barrier.

4.7 THE SCANNING TUNNELLING MICROSCOPE (STM)

The scanning tunnelling microscope was invented by Binnig and Rohrer in 1981 and provides extremely high resolution images of the <u>surfaces</u> of the materials under examination. It relies on the quantum mechanical phenomenon of tunnelling (see section 4.6) and therefore on the wave nature of electrons.

A <u>conducting</u> probe with an incredibly sharp tip (often only one atom across) is positioned between 0.1 nm and 1 nm above the surface being studied (Fig. 4.9). When a suitable PD ($\sim$ 1 V or less) is applied between the surface (which must be conducting) and the tip, electrons can tunnel through the potential energy barrier that exists in the gap between the surface and the tip, creating a small current. To a good approximation, the current is an <u>exponential</u> function of the width of the barrier, and this is approximately equal to the width of the gap. The tunnelling current therefore depends very critically on the height of the probe above the surface.

The probe is scanned across the surface whilst at the same time (in one mode of operation) a feedback system moves it up and down in such a way as to keep the tunnelling current constant. The probe is therefore maintained at a constant

Fig. 4.9
The scanning tunnelling
microscope

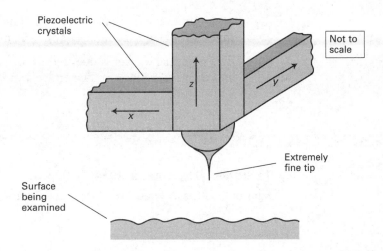

height above the surface and so traces out its profile. A large number of closely
spaced, parallel scans are made. The x, y and z coordinates of the tip are monitored
throughout and are processed using special software to produce a 'map' of the
surface on a VDU. Some examples are shown in Fig. 4.10.

Fig. 4.10
STM images: (a) Carbon
atoms in a sample of
highly orientated
pyrolytic graphite
(HOPG). The scale of this
image is measured in
angstroms (labelled 'A'
on the image;
1 angstrom = 0.1 nm)
and the imaging current
in nanoamperes (labelled
'na').
(b) Natural defect in
graphite surface.
(c) The atomically sharp
interface between GaAs
(upper left) and AlGaAs
(lower right) showing
individual arsenic atoms.
The area displayed
measures only 12 nm by
12 nm.

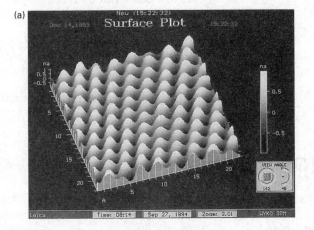

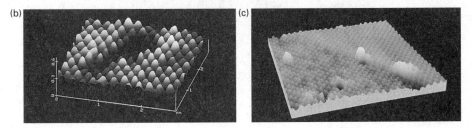

Notes

(i) The lateral resolution depends on the sharpness of the tip and is typically about 0.2 nm – good enough to reveal the positions of individual atoms.

(ii) The tunnelling current depends so critically on the distance between the probe and the sample surface that the heights of surface features can be measured to better than 10^{-3} nm!

(iii) The position of the probe (both lateral and vertical) is controlled by means of three mutually perpendicular piezoelectric crystals (crystals that change size when a PD is applied to them).

(iv)   Non-conducting surfaces can be examined if they are first coated with a thin layer of some suitable conducting material.

(v)   The potential barrier is due to the electrostatic force between an electron and the positive ions in the surface and the tip.

CONSOLIDATION

De Broglie's Hypothesis (1924)

A particle of mass m and velocity v has an associated wavelength λ where

$$\lambda = \frac{h}{p} = \frac{h}{mv}$$

This was confirmed by Davisson and Germer in 1927 when they diffracted electrons.

The square of the amplitude of the wave at any point is equal to the probability of the associated particle being at that point.

For an electron accelerated through a potential difference V

$$\lambda = \frac{h}{\sqrt{2meV}}$$

Double-Slit Experiment

Each electron passes through both slits and interferes with itself.

QUESTIONS ON CHAPTER 4

1. Calculate the de Broglie wavelength of an electron which has been accelerated from rest through a PD of 40 V.
   (Electron charge $= 1.6 \times 10^{-19}$ C, electron mass $= 9.1 \times 10^{-31}$ kg, Planck's constant $= 6.6 \times 10^{-34}$ J s.)

2. Calculate the de Broglie wavelength of (a) a person of mass 70 kg moving at $4.0 \, \text{km h}^{-1}$, (b) a car of mass 8.0×10^2 kg moving at $90 \, \text{km h}^{-1}$.
   (Planck's constant $= 6.6 \times 10^{-34}$ J s.)

3. Find an expression for the de Broglie wavelength of an electron in terms of its kinetic energy E, the electron mass m_e, and the Planck constant h.   [C]

4. What is the speed of an electron that has a de Broglie wavelength of 1.0×10^{-10} m?
   (Planck's constant $= 6.6 \times 10^{-34}$ J s, mass of electron $= 9.1 \times 10^{-31}$ kg.)

5. What is the ratio of the de Broglie wavelength of an electron to that of a proton (a) if they both have the same energy, (b) if they both have the same speed?
   (Mass of electron $= 9.1 \times 10^{-31}$ kg, mass of proton $= 1.7 \times 10^{-27}$ kg.)

6. (a) Electrons are known to show wave properties, as illustrated by electron diffraction.
   (i) When electrons are accelerated from rest by a potential difference, V, to a speed, v, in a vacuum, these quantities are related by the equation

   $$\frac{1}{2}mv^2 = eV$$

   where e is the charge of the electron and m is its mass.
   Explain how the equation is an application of the principle of conservation of energy.

(ii) Calculate the speed of electrons accelerated from rest through a PD of 2000 V.

(b) In a laboratory demonstration of electron diffraction, electrons are accelerated in a vacuum tube and pass through a thin disc of graphite on to a fluorescent screen at the end of the tube, where a pattern of concentric rings is seen.

(i) Use the momentum–wavelength equation to show that the wavelength associated with the electrons referred to in (a) (ii) above is 2.7×10^{-11} m.

(ii) By reference to your answer in (b) (i), explain why electrons can be diffracted through quite large angles by passing them through a thin sheet of graphite.

(iii) Describe one simple test you could carry out using the above apparatus to support the idea that these rings are produced by beams of *negatively-charged* particles. Explain, with the aid of a diagram, how you would reach this conclusion from the observations made.

(Charge on electron $= 1.6 \times 10^{-19}$ C, mass of electron $= 9.1 \times 10^{-31}$ kg, Planck's constant $= 6.6 \times 10^{-34}$ J s.)

[J, '94]

7. (a) Estimate the de Broglie wavelength of an electron that has been emitted thermionically in a vacuum from a filament and then accelerated through a PD of 30.0 kV.

(b) Outline the principle of operation of a transmission electron microscope (TEM).

(c) State one reason why a transmission electron microscope in operation is unable to achieve its theoretical resolving power.
($e = 1.60 \times 10^{-19}$ C,
$m_e = 9.11 \times 10^{-31}$ kg,
$h = 6.63 \times 10^{-34}$ J s.)

[J (specimen), '96]

8. Give a brief account of the quantum mechanical phenomenon of tunnelling.

9. Outline the principle of operation of the scanning tunnelling microscope.

5

ATOMIC ENERGY LEVELS

5.1 THE BOHR MODEL OF THE ATOM

In 1909 Geiger and Marsden, under the direction of Ernest Rutherford, investigated the scattering of α-particles by thin films of heavy metals*. The results of these experiments led Rutherford to propose that **an atom has a positively charged core (now called the nucleus) which contains most of the mass of the atom and which is surrounded by orbiting electrons**.

Although Rutherford's model of the atom was completely consistent with the experimental data, there was considerable opposition to it on theoretical grounds. An orbiting electron is constantly changing its direction and therefore is accelerating. According to classical electromagnetic theory (see section 2.4), when a charged particle is accelerated it emits electromagnetic radiation. The orbiting electrons, therefore, could be expected to emit radiation continuously. This is not possible, for if an electron were to emit radiation, it would have to do so at the expense of its own energy and as a consequence it would slow down and spiral into the nucleus, in which case the atom would cease to exist.

The problem was resolved by Neils Bohr in 1913. Bohr assumed that each electron moves in a circular orbit which is centred on the nucleus, the necessary centripetal force being provided by the electrostatic force of attraction between the positively charged nucleus and the negatively charged electron. On this basis he was able to show that the energy of an orbiting electron depends on (among other factors) the radius of its orbit. This much was obvious, but Bohr made two revolutionary proposals.

(i) The angular momenta of the electrons are whole-number multiples of $h/2\pi$, where h is a constant known as **Planck's constant**. (Thus, the angular momentum does not have a <u>continuous</u> range of values, i.e. it is **quantized**.) This means that the electrons can have only certain orbital radii, which in turn means that **the electrons are allowed to have only certain values of energy (called energy levels)**. This nullifies the idea that the electrons should continuously emit radiation, for if they were to do so, they would lose energy continuously and would need to have a <u>continuous</u> range of energies available to them. The allowed energy levels are often referred to as **stationary states**, since an electron can remain in an energy level indefinitely without radiating any energy.

(ii) An electron can jump from an orbit in which its energy is E_2, say, to one which is closer to the nucleus and of lower energy, E_1 say. In doing so, the electron gives up the energy difference of the two levels by emitting an electromagnetic wave whose frequency, f, is given by

*See, for example, R. Muncaster, *A-Level Physics* (Stanley Thornes).

$$E_2 - E_1 = hf \qquad [5.1]$$

where h = Planck's constant = 6.626×10^{-34} Js.

The principal justification for the Bohr model is that it predicts, to a high degree of accuracy, the wavelengths emitted by atomic hydrogen. It also works for other one-electron systems, such as singly ionized helium and doubly ionized lithium. However, it cannot explain the spectra of multi-electron atoms, nor can it account for the relative intensities of the various spectral lines. Another objection is that the model involves the <u>arbitrary</u> assumption that the allowed values of angular momentum are integral multiples of $h/2\pi$. (Over ten years later it was shown that electrons can be regarded as waves, and that the allowed values of angular momentum were consistent with the allowed orbits being exactly the right size to accommodate a stationary (standing) electron wave.)

The current model of the atom is based on **wave mechanics**. The electrons are no longer considered to move in definite orbits, but exist as an 'electron cloud' throughout the volume of the atom – the Bohr radii are the most probable positions of the electrons. The concepts of discrete energy levels and transitions between them giving rise to the emission or absorption of radiation are retained. The existence of discrete energy levels (the energies of which are the same as those given by Bohr) is a natural consequence of the theory and there is no need to make any arbitrary assumptions.

5.2 MATHEMATICAL TREATMENT OF THE HYDROGEN ATOM ACCORDING TO THE BOHR MODEL

Consider an electron of mass m and charge e moving with velocity v in a circular orbit of radius r about a hydrogen nucleus.* The charge on the nucleus is also e.

There is an inward directed Coulomb force acting on the electron, the magnitude of which, F, is given by

$$F = \frac{1}{4\pi\varepsilon_0}\frac{e^2}{r^2}$$

The centripetal acceleration of the electron is v^2/r. Therefore, by Newton's second law

$$F = m\frac{v^2}{r}$$

Combining these equations gives

$$\frac{e^2}{4\pi\varepsilon_0 r^2} = \frac{mv^2}{r} \qquad [5.2]$$

Multiplying each side of equation [5.2] by mr^3 gives

$$\frac{me^2 r}{4\pi\varepsilon_0} = (mvr)^2 \qquad [5.3]$$

*A hydrogen nucleus is much more massive than an electron and therefore it is a sufficiently good approximation to assume that the electron rotates about the nucleus, rather than about the centre of mass of the system.

According to Bohr's first proposal, the angular momentum mvr is given by

$$mvr = \frac{nh}{2\pi} \qquad (n = 1, 2, 3, \ldots)$$

Therefore, from equation [5.3]

$$\frac{me^2 r}{4\pi\varepsilon_0} = \left(\frac{nh}{2\pi}\right)^2$$

i.e. $\quad r = \dfrac{n^2 h^2 \varepsilon_0}{\pi m e^2}$ [5.4]

The total energy, E, of the system is given by

$$E = E_k + E_p \tag*{[5.5]}$$

where

E_k = the kinetic energy of the electron

$$= \tfrac{1}{2}mv^2 = \frac{1}{2}\frac{e^2}{4\pi\varepsilon_0 r} \qquad \text{(from equation [5.2])}$$

and

E_p = the potential energy of the electron

If the nucleus is considered to be a point charge, the electric potential at a distance r is given by

$$\frac{e}{4\pi\varepsilon_0 r}$$

Therefore, the work done in bringing an electron from infinity to a point a distance r from the nucleus is given by

$$-\frac{e^2}{4\pi\varepsilon_0 r}$$

(The minus sign arises because the nucleus attracts the electron.)

If the potential energy of the electron is taken to be zero when it is at infinity, then

$$E_p = \frac{-e^2}{4\pi\varepsilon_0 r}$$

Substituting for E_k and E_p in equation [5.5] gives

$$E = \frac{1}{2}\frac{e^2}{4\pi\varepsilon_0 r} - \frac{e^2}{4\pi\varepsilon_0 r}$$

i.e. $\quad E = -\dfrac{e^2}{8\pi\varepsilon_0 r}$

Therefore, from equation [5.4]

$$E = -\frac{me^4}{8\varepsilon_0^2 h^2}\cdot\frac{1}{n^2} \qquad (n = 1, 2, 3, \ldots) \tag*{[5.6]}$$

Notes   (i)   The energy is always negative (equation [5.6]). Work has to be done to remove the electron to infinity, where it is considered to have zero energy, i.e. the electron is 'bound' to the atom.

         (ii)   Increasing values of r are associated with increasing values of n and therefore with increasing (i.e. less negative) values of E.

5.3 ENERGY LEVELS

The energies of the electrons in an atom can have only certain values. These values are called the energy levels of the atom. All atoms of a given element have the same set of energy levels and these are characteristic of the element, i.e. they are different from those of every other element. The energies of the various levels can be found by measuring excitation potentials (see section 5.7) or can be calculated by using wave mechanics (and, in the specific case of the hydrogen atom, by using the Bohr model). It is convenient to express energy level values in **electronvolts** (see section 3.6), where one electronvolt (i.e. 1 eV) is equal to 1.6×10^{-19} joules.

The energy levels of an atom are usually represented as a series of horizontal lines. Fig. 5.1 shows the energy level diagram of the hydrogen atom. Hydrogen has only

Fig. 5.1
Energy levels of the
hydrogen atom

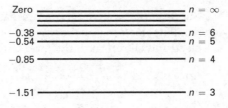

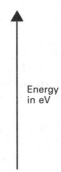

one electron and this normally occupies the lowest level and has an energy of $-13.6\,\mathrm{eV}$. When the electron is in this level the atom is said to be in its **ground state**. If the atom absorbs energy in some way (for example by being involved in a collision or by absorbing electromagnetic radiation), the electron may be promoted into one of the higher energy levels. (On the Bohr model the electron has moved into an orbit which is farther away from the nucleus.) The atom is now unstable – it is said to be in an **excited state** – and after a short, but random, interval the electron 'falls' back into the lowest level so that the atom returns to its ground state. The energy that was originally absorbed is emitted as an electromagnetic wave.

Each energy level is characterized by what is called a **quantum number,** n (see Fig. 5.1). The lowest level has $n = 1$, the next has $n = 2$, etc. The energy of the level which has $n = \infty$ is zero. If the electron is raised to this level, it becomes free of the atom. An atom which has lost an electron is said to be **ionized**, and therefore the energy required to ionize a hydrogen atom which is in its ground state is $13.6\,\mathrm{eV}$.

5.4 THE OPTICAL LINE SPECTRUM OF ATOMIC HYDROGEN

Suppose that a hydrogen atom has acquired energy in some way, and that as a result its electron is in the energy level characterized by $n = 4$. After a short time the electron will return to the level with $n = 1$. There are four possible 'routes'.

(a) $n = 4 \longrightarrow n = 3 \longrightarrow n = 2 \longrightarrow n = 1$

(b) $n = 4 \longrightarrow n = 3 \longrightarrow n = 1$

(c) $n = 4 \longrightarrow n = 2 \longrightarrow n = 1$

(d) $n = 4 \longrightarrow n = 1$

This involves six different transitions, namely:

$$n = 4 \longrightarrow n = 3 \qquad n = 4 \longrightarrow n = 2$$
$$n = 4 \longrightarrow n = 1 \qquad n = 3 \longrightarrow n = 2$$
$$n = 3 \longrightarrow n = 1 \qquad n = 2 \longrightarrow n = 1$$

($n = 4 \to n = 3$ appears in both (a) and (b), $n = 2 \to n = 1$ appears in both (a) and (c)). Each transition involves the emission of an electromagnetic wave whose frequency depends on the difference in energy of the two levels involved. When there are large numbers of atoms the different transitions take place simultaneously and radiation of many different frequencies is emitted. The line spectrum of hydrogen is composed of light of these frequencies.

When an electron moves from a level with energy E_2 to one of lower energy, E_1, the frequency, f, of the emitted radiation is given by

$$E_2 - E_1 = hf \tag{5.7}$$

where h = Planck's constant. When using this equation it is necessary to bear in mind that h and f are normally expressed in the relevant SI units ($\mathrm{J\,s}$ and Hz respectively), in which case E_1 and E_2 must be expressed in joules.

Suppose it is required to calculate the frequency, f, and the wavelength, λ, of the radiation emitted as a result of an electron transition from $n = 4$ to $n = 3$. From Fig. 5.1 the energies involved are $-0.85\,\text{eV}$ and $-1.51\,\text{eV}$, i.e.

$$E_2 \;=\; -0.85\,\text{eV} \;=\; -0.85 \times 1.6 \times 10^{-19} \;=\; -1.36 \times 10^{-19}\,\text{J}$$

and

$$E_1 \;=\; -1.51\,\text{eV} \;=\; -1.51 \times 1.6 \times 10^{-19} \;=\; -2.42 \times 10^{-19}\,\text{J}$$

If the value of h is taken to be $6.6 \times 10^{-34}\,\text{J s}$, then by equation [5.7]

$$(-1.36 \times 10^{-19}) - (-2.42 \times 10^{-19}) \;=\; 6.6 \times 10^{-34} \times f$$

i.e. $\quad 1.06 \times 10^{-19} \;=\; 6.6 \times 10^{-34} \times f$

i.e. $\quad f \;=\; \dfrac{1.06 \times 10^{-19}}{6.6 \times 10^{-34}}$

i.e. $\quad f \;=\; 1.6 \times 10^{14}\,\text{Hz}$

Fig. 5.2
The main spectral transitions of atomic hydrogen

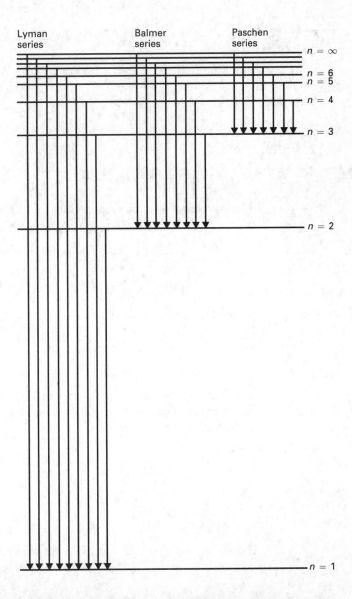

Also, $\lambda = c/f$, where c = the velocity of light = $3 \times 10^8 \, \text{m s}^{-1}$. Therefore

$$\lambda = \frac{3 \times 10^8}{1.6 \times 10^{14}}$$

i.e. $\quad \lambda = 1.9 \times 10^{-6} \, \text{m}$

Wavelengths calculated in this way are in excellent agreement with those observed in the line spectrum of atomic hydrogen – providing convincing evidence of the existence of energy levels.

The spectrum of atomic hydrogen contains distinct groups of lines. The three most obvious groups are the Lyman series, the Balmer series and the Paschen series. The wavelengths of the lines in the Lyman series are in the ultraviolet and each is associated with a transition involving the level with $n = 1$ (see Fig. 5.2). The Balmer series involves transitions to the level with $n = 2$, and as a consequence smaller energy differences are involved and the wavelengths are in the visible. The lines of the Paschen series are in the infrared.

QUESTIONS 5A

1. Refer to Fig. 5.1. Calculate the frequency and wavelength of the radiation resulting from the following transitions: **(a)** $n = 4$ to $n = 2$, **(b)** $n = 2$ to $n = 1$. ($h = 6.6 \times 10^{-34} \, \text{J s}$, $c = 3.0 \times 10^8 \, \text{m s}^{-1}$.)

5.5 THE RYDBERG FORMULA

The wavelengths of the various lines in the spectrum of atomic hydrogen are given by **the Rydberg formula**

$$\frac{1}{\lambda} = R_\text{H} \left(\frac{1}{n_1{}^2} - \frac{1}{n_2{}^2} \right) \quad \begin{pmatrix} n_2 > n_1 \\ n_1 = 1, 2, 3, \ldots \\ n_2 = 2, 3, 4, \ldots \end{pmatrix}$$

where R_H is a constant called **the Rydberg constant for hydrogen**. (For the Lyman series $n_1 = 1$, $n_2 = 2, 3, 4, \ldots$; for the Balmer series $n_1 = 2$, $n_2 = 3, 4, 5, \ldots$, etc.) From spectroscopic data

$$R_\text{H} = 1.0968 \times 10^7 \, \text{m}^{-1}$$

The Rydberg formula is easily explained on the basis of the Bohr model. If E_1 and E_2 are the energies of two levels with quantum numbers n_1 and n_2 respectively, then from equation [5.6]

$$E_1 = -\frac{me^4}{8\varepsilon_0{}^2 h^2} \cdot \frac{1}{n_1{}^2}$$

and

$$E_2 = -\frac{me^4}{8\varepsilon_0{}^2 h^2} \cdot \frac{1}{n_2{}^2}$$

If $E_2 > E_1$, then since $f = c/\lambda$, the wavelength, λ, of the radiation due to a transition from E_2 to E_1 is given by equation [5.7] as

$$\frac{hc}{\lambda} = E_2 - E_1$$

i.e. $$\frac{hc}{\lambda} = \frac{me^4}{8\varepsilon_0^2 h^2} \left(\frac{1}{n_1^2} - \frac{1}{n_2^2} \right)$$

i.e. $$\frac{1}{\lambda} = \frac{me^4}{8\varepsilon_0^2 h^3 c} \left(\frac{1}{n_1^2} - \frac{1}{n_2^2} \right)$$

which we may write as

$$\frac{1}{\lambda} = R_\infty \left(\frac{1}{n_1^2} - \frac{1}{n_2^2} \right) \qquad [5.8]$$

where

$$R_\infty = \frac{me^4}{8\varepsilon_0^2 h^3 c}$$

Using known values of the constants on the right-hand side of this expression gives

$$R_\infty = 1.0974 \times 10^7 \, \text{m}^{-1}$$

which is extremely close to the value of R_H given previously.

The slight difference between R_H and R_∞ is accounted for by making a minor refinement to the Bohr theory. In deriving equation [5.6], like Bohr himself, we have assumed that the electron rotates about the nucleus. In truth, both the electron and the nucleus rotate about their common centre of mass. When this is taken into account m in equation [5.6] has to be replaced by the so-called **reduced mass** $mM/(m+M)$ where M is the mass of the hydrogen nucleus. Equation [5.8] then becomes

$$\frac{1}{\lambda} = \left(\frac{M}{m+M} \right) R_\infty \left(\frac{1}{n_1^2} - \frac{1}{n_2^2} \right)$$

Since $M = 1836m$, we find

$$\frac{M}{m+M} R_\infty = 1.0968 \times 10^7 \, \text{m}^{-1}$$

which is the same as the value of R_H.

Note The symbol R_∞ is used because equation [5.8] gives the wavelength we would expect to find for a nucleus of <u>infinite</u> mass.

5.6 IONIZATION AND EXCITATION POTENTIALS

The <u>minimum</u> amount of energy required to ionize an atom which is in its ground state, i.e. to remove its most loosely bound electron, is called the **first (or principal) ionization energy** of the atom. For example, the first ionization energy of hydrogen is 13.6 eV, i.e. 2.18×10^{-18} J (see Fig. 5.1). The energy to remove the next most <u>loosely</u> bound electron is called the second ionization energy, etc. The first <u>ionization</u> energy is often referred to simply as **the ionization energy**. This is particularly so in the case of hydrogen which has only one electron and therefore no higher ionization energies than the first.

It follows from the definition of the electronvolt that 13.6 eV is the kinetic energy gained by an electron in being accelerated through a PD of 13.6 V. Therefore, if an electron which has been accelerated from rest by a PD of 13.6 V collides with a hydrogen atom, it has exactly the right amount of energy to produce ionization. This is a common method of producing ionization, and therefore the term ionization <u>potential</u> is often used. **Ionization potential** is expressed in <u>volts</u> and is <u>numerically</u> equal to the ionization energy. Thus, the ionization potential of hydrogen is 13.6 V.

The energy required to excite an atom which is in its ground state is called an **excitation energy** of the atom. For example, the first and second excitation energies of hydrogen are (approximately) 10.2 eV and 12.1 eV respectively (see Fig. 5.1). The corresponding **excitation potentials** are 10.2 V and 12.1 V.

5.7 MEASUREMENT OF EXCITATION POTENTIALS BY ELECTRON COLLISION

The spectroscopic evidence for the existence of discrete energy levels (see section 5.4) is supported by the results of experiments in which electrons are caused to collide with gas atoms. The first successful experiments of this type were carried out by Franck and Hertz in 1914. A schematic form of the apparatus used in a Franck–Hertz type experiment is shown in Fig. 5.3.

Fig. 5.3
Apparatus for a Franck–
Hertz type experiment

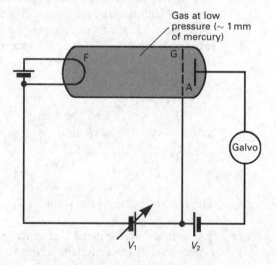

Electrons which have been emitted thermionically by the filament (F) are accelerated towards the grid (G) and pass through it. The anode (A) is <u>slightly</u> <u>negative</u> with respect to G and therefore the electrons are retarded as they move from G to A. If V_1 is slightly larger than V_2, the electrons have sufficient energy to reach A and a current flows through the galvanometer. If V_1 is increased, the current increases at first (see Fig. 5.4). As the electrons move across the tube they collide with gas atoms, and at this stage the collisions are elastic – the electrons bounce off the atoms without losing any energy. As V_1 is increased further, eventually a point is reached when the electrons have exactly the right amount of energy to promote the atomic electrons to higher energy levels and <u>inelastic</u> collisions result. All the energy of a bombarding electron is given up to the atom with which it has collided. The electrons no longer have sufficient energy to reach A and the galvanometer current falls. By determining the value of V_1 at which this

Fig. 5.4
Typical results of a
Franck–Hertz type
experiment

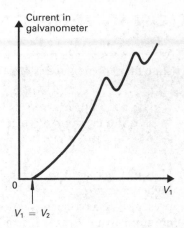

happens it is possible to calculate the excitation potential of the transition which
has taken place. Increasing V_1 beyond this value causes the current to rise until V_1
is again such that the bombarding electrons can produce excitation, whereupon
the current falls.

5.8 SPONTANEOUS AND STIMULATED EMISSION

An atom in an energy state E_1 can be raised to a higher state E_2 by absorbing a
photon whose energy is exactly equal to $E_2 - E_1$ (Fig. 5.5(a)). The process is
known as **stimulated absorption**.

In the process known as **spontaneous emission** (Fig. 5.5(b)) an atom which is
already in some excited state E_2 spontaneously drops down to a lower state E_1 by
emitting a photon of energy $E_2 - E_1$.

Fig. 5.5
(a) Stimulated absorption,
(b) Spontaneous emission,
(c) Stimulated emission

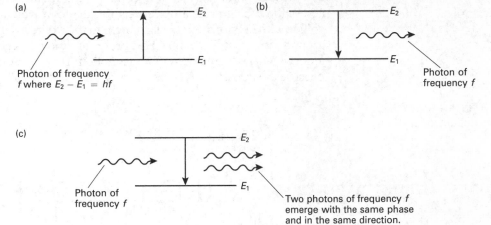

In 1917 Einstein showed (on theoretical grounds) that an atom in an excited state
E_2 can be triggered into undergoing a transition to a lower state E_1 as a result of
encountering a photon of energy $E_2 - E_1$ (Fig. 5.5(c)). This is known as
stimulated emission and the atom emits a photon which not only has the same
frequency as the incident photon, but which also has the same phase and moves in
the same direction.

5.9 POPULATION INVERSION AND LASER ACTION

When a system is in thermal equilibrium there are fewer atoms in any high-energy state than in any lower state.

> The term **population inversion** is used to describe a (non-equilibrium) situation in which there are <u>more</u> atoms in a high-energy state than in some lower state.

Consider a system in which there are n_1 atoms of energy E_1 and n_2 of some <u>higher</u> energy E_2. Suppose that a photon of frequency f, where $E_2 - E_1 = hf$, is present. We have seen (section 5.8) that this photon may be absorbed by raising an atom from E_1 to E_2, or it may produce an identical photon through the process of stimulated emission by triggering the transition E_2 to E_1. It can be shown that

$$\frac{\text{Probability of absorption}}{\text{Probability of stimulated emission}} = \frac{n_1}{n_2}$$

If we could create a situation in which $n_2 > n_1$ (i.e. in which there is a population inversion), the probability of stimulated emission would exceed that of absorption and a second photon would be likely to be produced. A chain reaction would ensue, producing large numbers of photons all travelling in the same direction and all with the same phase as each other, i.e. a beam of high-intensity <u>coherent</u> radiation would be produced. This is what happens in a laser.

5.10 THE HELIUM–NEON GAS LASER

A mixture of helium and neon (90% He, 10% Ne) is contained in a discharge tube at a pressure of about 1 mmHg (Fig. 5.6). A radio-frequency supply connected across the tube creates an electrical discharge in which (in particular) electrons collide with helium atoms raising them to the excited state E_1, 20.61 eV above the ground state, E_0 (Fig. 5.7). (This process is known as **electrical pumping**.) An atom which has been raised to E_1 cannot return to E_0 simply by emitting a photon, because the transition $E_1 \rightarrow E_0$ is quantum-mechanically forbidden* – the atom is

Fig. 5.6
Helium–neon laser

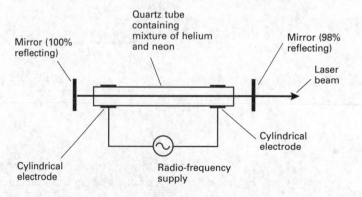

*It follows from quantum-mechanical considerations beyond the scope of this book that transitions between energy levels can occur only if certain conditions are fulfilled; transitions that cannot occur are said to be **quantum-mechanically forbidden**.

Fig. 5.7
Energy level diagram for
helium–neon laser

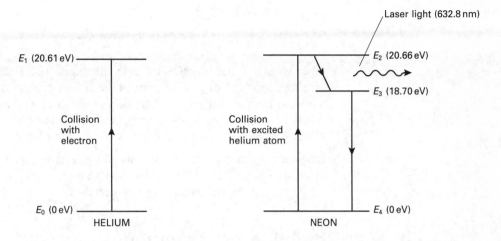

said to be in a **metastable state**. However, if it collides with a neon atom which is in its ground state (E_4), it may transfer its energy to the neon by raising it to the excited state, E_2, 20.66 eV above E_4. (The extra 0.05 eV required is supplied by the kinetic energy of the helium atom.) This creates a situation in which there are more neon atoms in E_2 then there are in the lower state, E_3, i.e. **there is a population inversion between states E_2 and E_3**.

Spontaneous emission between E_2 and E_3 produces photons which, because of the population inversion, are more likely to stimulate the emission of other photons than they are to be absorbed. The majority of the photons which are moving parallel to the axis of the tube are reflected back and forth, many times, by the mirrors at the ends of the tube, stimulating the emission of more and more axial photons, thus creating a high-intensity beam of radiation, a proportion of which emerges through the partially transmitting mirror.

Notes

(i) Spontaneous emission between E_3 and E_4 occurs more rapidly than that between E_2 and E_3 and therefore enhances the population inversion between E_2 and E_3.

(ii) Neon atoms are also excited to E_2 through collisions with electrons.

(iii) The name 'laser' is an acronym for **light amplification by the stimulated emission of radiation**.

(iv) Helium–neon lasers normally produce red light (632.8 nm) but some are designed to give yellow or green.

5.11 PROPERTIES OF LASER LIGHT

Coherent

Laser light is coherent because all the photons in a beam of laser light are produced by stimulated emission and are therefore all in phase with each other. In contrast, ordinary light (e.g. light from a filament lamp or a sodium vapour lamp) is incoherent because it is produced by the random process of spontaneous emission.

Unidirectional

The beam of light produced by a laser has very little divergence (typically $1'$ of arc). Photons which are not moving parallel (or very nearly parallel) to the axis of the tube can undergo no more than a few reflections before they are directed out of the tube through the side walls. Bearing in mind that a photon produced by stimulated emission moves in the same direction as the photon that stimulated it, it follows that only those photons which are travelling more or less parallel to the axis of the tube can build up in numbers to any significant extent.

Conventional light sources emit in all directions. Attempts to produce parallel beams involve the use of lenses and mirrors, and such beams do not come anywhere near matching the highly directional quality of laser light.

Highly monochromatic

It is impossible for light (or any other form of electromagnetic radiation) to be truly monochromatic. (This is a consequence of Heisenberg's uncertainty principle – see note below.) However, the light produced by lasers is very nearly so. That from a helium–neon laser, for example, has a wavelength spread (i.e. a **linewidth**) of only about 10^{-6} nm – over a hundred times narrower than the best that can be achieved with an ordinary discharge tube.

Note   According to the **Heisenberg uncertainty principle**, the energy of a system that spends a time Δt in any particular state is uncertain to an extent ΔE where

$$\Delta E \, \Delta t \approx h/2\pi$$

Since the time that an atom spends in an excited state before undergoing a transition to some lower state is finite (typically $\sim 10^{-8}$ s), it follows that $\Delta E > 0$, and therefore there is an uncertainty in the energy of the photon emitted as a result of the transition. Since $E = hf = hc/\lambda$, there are corresponding uncertainties in the frequency and wavelength of the photon. It follows that the associated spectral line extends over a range of frequencies and wavelengths. This spread is unavoidable and is known as the **natural line width** of the line.

CONSOLIDATION

Bohr Model of Atom

Electrons are allowed to have only certain orbital radii and therefore only certain values of energy (called energy levels).

An electron may 'jump' from a level with energy E_2 to a lower level of energy E_1 causing the emission of electromagnetic radiation of frequency f where

$$E_2 - E_1 \; = \; hf$$

The principal justification for the Bohr model is that it predicts, to a high degree of accuracy, the wavelengths emitted by atomic hydrogen. However, it cannot explain the relative intensities of the various spectral lines, nor can it account for the spectra of multi-electron atoms.

According to the Bohr model

$$E \; = \; -\frac{me^4}{8\varepsilon_0{}^2 h^2} \cdot \frac{1}{n^2} \qquad \text{(for hydrogen)}$$

which accounts for the **Rydberg formula**

$$\frac{1}{\lambda} = R_H \left(\frac{1}{n_1^2} - \frac{1}{n_2^2} \right) \qquad (n_2 > n_1)$$

A photon produced by **stimulated emission** has the same frequency, phase and direction as the incident photon.

Lasers produce light by the process of stimulated emission and require a non-equilibrium condition called a **population inversion**, i.e. more atoms in a high energy state than in some lower state.

Laser light is coherent, unidirectional and highly monochromatic.

QUESTIONS ON CHAPTER 5

1. The figure below representing the lowest energy levels of the electron in the hydrogen atom, gives the principal quantum number n associated with each, and the corresponding value of the energy, measured in joules.

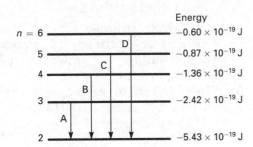

(a) Calculate the wavelengths of the lines arising from the transitions marked A, B, C, D on the figure.

(b) Show that the other transitions that can occur give rise to lines which are in either the ultraviolet or the infrared regions of the spectrum.

(c) The level $n = 1$ is the 'ground state' of the unexcited hydrogen atom. Explain why hydrogen in its ground state is quite transparent to light emitted by the transitions A, B, C, D, and also what happens when 21.7×10^{-19} J of energy is supplied to a hydrogen atom in its ground state.

(Take the value of the speed of light in vacuum, c, to be 3.00×10^8 m s^{-1}, and that of the Planck constant, h, to be 6.63×10^{-34} J s.) [O]

2. (a) Calculate the energy of one photon of light emitted within the D-lines of a sodium lamp if the wavelength of the D-lines is 589 nm.

(b) In a 200 W sodium street lamp, 30% of input electrical energy is emitted within the D-lines. How many photons of light are emitted within the D-lines per second?
$(h = 6.6 \times 10^{-34}$ J s, $c = 3.0 \times 10^8$ m s^{-1}.)
[W, '91]

3. The ionisation energy for a hydrogen atom is 13.6 eV if the atom is in its ground state. It is 3.4 eV if the atom is in the first excited state.
Explain the terms *ionisation energy* and *excited state*.
Calculate the wavelength of the photon emitted when a hydrogen atom returns to the ground state from the first excited state. Name the part of the electromagnetic spectrum to which this wavelength belongs.
(Electronic charge, $e = -1.60 \times 10^{-19}$ C, the Planck constant, $h = 6.63 \times 10^{-34}$ J s, speed of light, $c = 3.00 \times 10^8$ m s^{-1}.) [L, '92]

4. The energy levels of the hydrogen atom are given by the expression
$$E_n = -2.16 \times 10^{-18}/n^2 \text{ J}$$
where n is an integer.
(a) What is the ionization energy of the atom?
(b) What is the wavelength of the H_α line, which arises from transitions between $n = 3$ and $n = 2$ levels?
(The Planck constant $= 6.6 \times 10^{-34}$ J s, the speed of electromagnetic radiation $= 3.0 \times 10^8$ m s^{-1}.) [S]

5. The lowest energy level in a helium atom (the ground state) is $-24.6\,\text{eV}$. There are a number of other energy levels, one of which is at $-21.4\,\text{eV}$.

(a) Define an eV.

(b) (i) Explain the significance of the negative signs in the values quoted.

(ii) What is the energy, in J, of a photon emitted when an electron returns to the ground state from the energy level at $-21.4\,\text{eV}$?

(iii) Calculate the wavelength of the radiation emitted in this transition.

The electronic charge, $e = 1.6 \times 10^{-19}\,\text{C}$. The speed of electromagnetic radiation, $c = 3.0 \times 10^{8}\,\text{m s}^{-1}$. The Planck constant, $h = 6.6 \times 10^{-34}\,\text{J s}$.

(c) Helium was first discovered from observations of the absorption spectrum produced by helium in the Sun's atmosphere.

What is an *absorption spectrum*?

[AEB, '89]

6. The frequencies f of the spectral lines emitted by atomic hydrogen can be represented by the expression

$$f = a\left(\frac{1}{n_1{}^2} - \frac{1}{n_2{}^2}\right)$$

where a is a constant and n_1 and n_2 are integers with $n_2 > n_1$.

Given that the ionization potential for atomic hydrogen is $13.6\,\text{V}$:

(a) Calculate the energies (in eV) of the quanta emitted by electronic transitions between the levels given by

(i) $n_1 = 2, n_2 = 3$
(ii) $n_1 = 1, n_2 = 2$.

(b) State in which parts of the spectrum these quanta are found.   [W, '92]

7. Hydrogen atoms in a discharge tube emit spectral lines whose frequencies f are given by

$$f = cR_{\text{H}}\left(\frac{1}{n_1{}^2} - \frac{1}{n_2{}^2}\right)$$

where $c = 3 \times 10^{8}\,\text{m s}^{-1}$, $R_{\text{H}} = 1.10 \times 10^{7}\,\text{m}^{-1}$ and n_1 and n_2 are any positive whole numbers.

(a) Calculate (i) the highest and (ii) the lowest frequencies in the Lyman series of spectral lines.

($n_1 = 1$ in the Lyman series.)

8. Some of the energy levels of the hydrogen atom are shown (not to scale) in the diagram.

Energy/eV   0.00 _____

          −0.54 _____

          −0.85 _____

          −1.51 _____

          −3.39 _____

          −13.58 _____ Ground state

$1\,\text{eV} = 1.6 \times 10^{-19}\,\text{J}$

(a) Why are the energy levels labelled with negative energies?

(b) State which transition will result in the emission of radiation of wavelength $487\,\text{nm}$. Justify your answer by suitable calculation.

(c) What is likely to happen to a beam of photons of energy (i) $12.07\,\text{eV}$, (ii) $5.25\,\text{eV}$, when passed through a vapour of atomic hydrogen?

($1\,\text{eV} = 1.6 \times 10^{-19}\,\text{J}$.)   [O & C, '92]

9. The energy levels of atomic hydrogen are given by

$$E_n = 13.6/n^2\,\text{eV} \quad (n = 1, 2, 3, \ldots)$$

A beam of $12.4\,\text{eV}$ electrons enters a gas which consists entirely of hydrogen atoms in the ground state.

(a) What frequencies of radiation are likely to be emitted?

(b) What frequencies would be emitted if the electrons were replaced by a beam of $12.4\,\text{eV}$ photons?

($e = 1.6 \times 10^{-19}\,\text{C}$, $h = 6.6 \times 10^{-34}\,\text{J s}$.)

10. (a) (i) Describe briefly the Bohr model for the hydrogen atom, and

(ii) state the important assumptions that Bohr made.

(b) Multiply each answer in (a) by h (Planck's constant $h = 6.6 \times 10^{-34}\,\text{J s}$) and state what each answer represents.   [W, '90]

(b) The table shows some of the energy levels for the hydrogen atom.

	Energy/eV
a	0
b	−0.54
c	−0.85
d	−1.51
e	−3.39
f	−13.6

(i) Define the electron volt.

(ii) Explain why the energy levels are given negative values.

(iii) How might the atom be changed from state e to state c?

(iv) State which level corresponds to the ground state.

(v) Calculate the ionization energy of atomic hydrogen in joules.

(vi) The result of the Bohr theory for the hydrogen atom can be expressed by

$$\frac{1}{\lambda} = R_\text{H}\left[\frac{1}{n_1{}^2} - \frac{1}{n_2{}^2}\right]$$

where n_1 and n_2 are whole numbers: for the ground state $n_1 = 1$.

Calculate the value of R_H.
($h = 6.6 \times 10^{-34}\,\text{J s}$; $e = -1.6 \times 10^{-19}\,\text{C}$;
$c = 3 \times 10^8\,\text{m s}^{-1}$.)   [W]

11. Calculate the number of photons per second emitted by a 1.2 mW helium–neon laser operating at 633 nm.
($h = 6.6 \times 10^{-34}\,\text{J s}$, $c = 3.0 \times 10^8\,\text{m s}^{-1}$.)

12. In what ways does the light produced by a laser differ from that produced by a filament lamp?

13. (a) What is meant by the terms *population inversion* and *stimulated emission*?

(b) Why is it necessary to create a population inversion in order to produce laser light?

6
THE THEORY OF SPECIAL RELATIVITY

6.1 THE ETHER

It used to be believed that there existed throughout the whole of space, and even inside matter itself, a medium known as **the ether**. There has never been any direct evidence for its existence, but it was considered necessary in order to account for the fact that light could travel through a vacuum. This was in the mistaken belief that because sound and all other (non-electromagnetic) wave motions were known to require a medium through which to travel, then so too must light. It was also supposed that **light travels at a fixed speed with respect to the ether**.

Since the ether was supposed to permeate the whole of space, it could be regarded as being a perfect frame of reference relative to which all motion could be measured and which therefore gave meaning to the concept of absolute motion.

6.2 THE MICHELSON–MORLEY EXPERIMENT (1887)

The reader should be familiar with the phenomenon of optical interference before continuing.

(a) Albert Abraham
    Michelson (1852–1931)
(b) Edward Williams
    Morley (1838–1923)

(a)

(b)

If the ether exists, the Earth must be moving through it as it orbits the Sun. A number of experiments have been carried out in attempts to detect this motion. The most famous (and at the time, the most sensitive) was that performed by Michelson and Morley in 1887. If the experiment had had its expected outcome, it would have:

(i)   confirmed the existence of the ether, and

(ii)   determined the <u>absolute</u> speed of the Earth through space.

Although it failed in both respects, its outcome was described by Bernal as 'the greatest null result in the history of science'.

The principle of the experiment was to compare the speed of light measured in the direction of the Earth's motion with that at right angles to it. The experiment made use of an instrument known as a Michelson interferometer (Fig. 6.1). Light from a monochromatic source at S is partly transmitted and partly reflected on reaching the half-silvered mirror, A. The two beams are then reflected by mirrors M_1 and M_2 respectively, so that they return to A. On reaching A the beams superpose and produce interference fringes which can be seen by an observer at O. (Some light also travels towards the source but this is of no consequence.)

Fig. 6.1
The interferometer used in the Michelson–Morley experiment

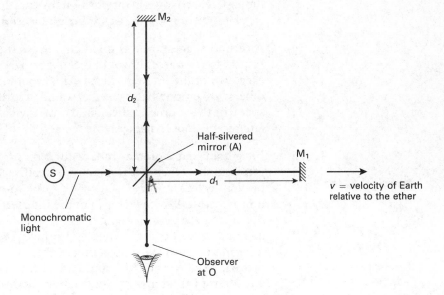

Suppose that the Earth (and therefore the interferometer) is moving to the right. Quite apart from any difference in the distances of M_1 and M_2 from A, because light travels at a fixed speed <u>with respect to the ether</u>, the time taken for light to travel from A to M_1 and back will not be the same as that from A to M_2 and back. The method involved rotating the interferometer through $90°$ (about a vertical axis) whilst observing the fringes produced at A. This was expected to cause the fringes to shift slightly because one path would now be shorter (in terms of time) and the other would be longer, and this would alter the phase difference between the two beams reaching A.

The pattern was expected to shift by an amount corresponding to just under 2/5 of a wavelength of sodium light. (Half a wavelength would cause each bright fringe to move to the position previously occupied by an adjacent dark fringe etc.) The observed shift corresponded to less than 1/100 of a wavelength, and this could be taken to be zero because there was a possible experimental error of 1/100 of a wavelength. The experiment was repeated on a number of occasions at different times of the year because it was possible that, at the time of the first measurement, the Sun itself was moving through space in just such a way that the Earth was at rest with respect to the ether. The result was always the same – **if the ether existed, the Earth appeared to be at rest relative to it.**

Implications of the Result

The Earth moves relative to the stars and to the Sun and to all the planets in the Solar System, and therefore the idea that the Earth really is at rest with respect to the ether can be dismissed immediately for it would make the Earth unique.

Three other ideas were put forward in attempts to explain the result of the Michelson–Morley experiment.

(i)   It was suggested by Michelson that the ether in the vicinity of the Earth might be dragged along by the Earth, in which case both arms of the interferometer would be at rest with respect to the ether. This was rejected on the grounds that it was in disagreement with observations concerning the apparent changes in the positions of stars due to the motion of the Earth. (A phenomenon known as **stellar aberration**.)

(ii)   A second suggestion was that light always travels at the same speed with respect to its source. It would therefore travel at the same speed relative to each arm of the interferometer. This would make light different from all other wave motions (the speed of sound, for example, is independent of the speed of the source of the sound). In any case, observations of the light emitted by double star systems seemed to rule out this possibility.

(iii) Fitzgerald and Lorentz, independently, proposed that objects moving through the ether with velocity v contract in the direction of their motion by a factor of $(1 - v^2/c^2)^{1/2}$ – the so-called **Fitzgerald–Lorentz contraction**. Whichever arm of the interferometer was moving through the ether parallel to its length would therefore contract by this amount and this would be just enough to produce the observed null result. There were many reasons for objecting to this hypothesis, but none that could prove it to be wrong. Nor could it be proved to be correct. For example, any attempt to verify the contraction by direct measurement would be bound to fail because whatever was used as a 'ruler' would itself contract!

We can draw the following conclusion:

The null result of the Michelson–Morley experiment allows just two possibilities. Either:

(i)   the ether does not exist, or

(ii)   it does exist but is impossible to detect because the Fitzgerald–Lorentz contraction makes it impossible to detect the motion of any object relative to it.

In either case we have to abandon the idea of absolute motion, and have to accept that all observers will obtain the same value for the speed of light regardless of their velocity.

Theory

Suppose the interferometer (Fig. 6.1) is moving to the right with speed v relative to the ether. If the speed of light relative to the ether is c, then the speed of light relative to the interferometer is $c - v$ from A to M_1 and is $c + v$ from M_1 to A. Therefore

$$\begin{array}{l}\text{Time for}\\\text{path A}\,M_1\,\text{A}\end{array} = \frac{d_1}{c - v} + \frac{d_1}{c + v} = \frac{2cd_1}{c^2 - v^2} = \frac{2d_1}{c(1 - v^2/c^2)}$$

$$\approx \frac{2d_1}{c}\left(1 + \frac{v^2}{c^2}\right)^*$$

Fig. 6.2
Diagram for theory of the Michelson–Morley experiment

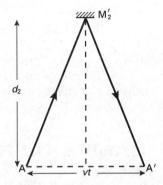

If light travels from A to M_2 and back in time t, the interferometer will move to the right by vt whilst this is happening, and the light path relative to the ether will be as shown in Fig. 6.2, where M_2' is the position of M_2 when the light reaches it and A′ is the position of A when the light returns to A. By Pythagoras' theorem

$$\left(AM_2'\right)^2 = d_2{}^2 + \left(\frac{vt}{2}\right)^2$$

$$\therefore \quad \left(\frac{ct}{2}\right)^2 = d_2{}^2 + \left(\frac{vt}{2}\right)^2$$

$$\therefore \quad \frac{t^2}{4}\left(c^2 - v^2\right) = d_2{}^2$$

Therefore

$$\begin{array}{l}\text{Time for}\\\text{path A}\,M_2\,\text{A}\end{array} = t = \frac{2d_2}{(c^2 - v^2)^{1/2}} = \frac{2d_2}{c(1 - v^2/c^2)^{1/2}} \approx \frac{2d_2}{c}\left(1 + \frac{v^2}{2c^2}\right)^*$$

$$\therefore \quad \text{Time difference} \approx \frac{2d_1}{c}\left(1 + \frac{v^2}{c^2}\right) - \frac{2d_2}{c}\left(1 + \frac{v^2}{2c^2}\right) = T_0 \ \text{ say}$$

If the apparatus is rotated through $90\,°$, then by analogy with what we have just done

$$\begin{array}{l}\text{Time for}\\\text{path A}\,M_1\,\text{A}\end{array} \approx \frac{2d_1}{c}\left(1 + \frac{v^2}{2c^2}\right)$$

$$\begin{array}{l}\text{Time for}\\\text{path A}\,M_2\,\text{A}\end{array} \approx \frac{2d_2}{c}\left(1 + \frac{v^2}{c^2}\right)$$

*This approximation follows from the binomial expansion.

$$\therefore \qquad \text{Time difference} \approx \frac{2d_1}{c}\left(1 + \frac{v^2}{2c^2}\right) - \frac{2d_2}{c^2}\left(1 + \frac{v^2}{c^2}\right) = T_{90} \text{ say}$$

$$\text{Change in time difference} \approx T_0 - T_{90}$$

$$\approx \frac{2d_1}{c}\left(\frac{v^2}{2c^2}\right) + \frac{2d_2}{c}\left(\frac{v^2}{2c^2}\right)$$

$$\approx \frac{(d_1 + d_2)\,v^2}{c^3}$$

$$\therefore \qquad \text{Change in number of wavelengths path difference} \approx \frac{c}{\lambda}\cdot\frac{(d_1 + d_2)\,v^2}{c^3}$$

$$\approx \frac{(d_1 + d_2)\,v^2}{\lambda c^2}$$

Putting $d_1 = d_2 = 11\,\text{m}$, $\lambda = 5.9 \times 10^{-7}\,\text{m}$ (the wavelength of sodium light), $c = 3.0 \times 10^8\,\text{m s}^{-1}$ and $v = 3.0 \times 10^4\,\text{m s}^{-1}$ (the mean orbital speed of the Earth) gives a change in path difference of 0.37 wavelengths.

6.3 INERTIAL FRAMES OF REFERENCE

A drinks trolley left unattended in the aisle of an aeroplane which is accelerating down the runway during take off would accelerate towards the back of the plane. This appears to contradict Newton's first law for there is no backward directed force acting on the trolley. Whilst the plane was stationary on the runway prior to take off, the trolley would also be stationary. This, of course, is in total accord with the first law – there is no force on the trolley and it remains at rest. Thus there are some reference frames (e.g. the stationary aeroplane) in which Newton's first law is valid and others (e.g. the accelerating aeroplane) in which it is not. As a second example, consider a particle moving from A to B just above a rotating turntable (Fig. 6.3). An observer at X sees the particle move in a straight line, whereas an observer on the turntable regards the path as being <u>curved</u>. The rotating turntable is therefore another example of a reference frame in which Newton's first law is not obeyed.

Fig. 6.3
Illustration of a non-inertial reference frame

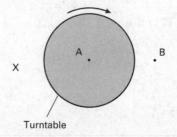

Turntable

> A reference frame in which Newton's first law is valid is called **an inertial frame of reference**.
>
> It can be shown that if any particular frame is an inertial frame, then any frame which is moving with constant velocity and which is not rotating relative to this frame is also an inertial frame.

Note   For most purposes it is a sufficiently good approximation to consider that the Earth is an inertial frame, despite the fact that it is rotating about the Sun.

6.4 THE POSTULATES OF THE THEORY OF SPECIAL RELATIVITY

The theory of special (or restricted) relativity, published by Einstein in 1905, is concerned only with inertial frames of reference. The theory of general relativity (1916) deals with the broader issue of non-uniform relative motion. We shall not be concerned with general relativity.

Albert Einstein
(1879–1955)

Special relativity is based on two postulates.

1. The laws of physics have the same form in all inertial frames of reference. (This is known as the **principle of special relativity**.)

2. The speed of light (in vacuum) is the same in all inertial frames; it does not depend on the velocity of either the source or the observer.

The first postulate implies that an experiment performed in one inertial frame will have exactly the same outcome when performed in any other inertial frame. It follows that no one frame can be singled out as being at rest and therefore that **the concept of absolute motion is meaningless, i.e. all motion is relative**. This is entirely consistent with Michelson's and Morley's failure to detect the motion of the Earth through the ether. The second postulate is also in agreement with the outcome of the Michelson–Morley experiment for it means that the speed of light measured in the direction of the Earth's motion is the same as that at 90° to it, and this is what they found.

These simple postulates have far-reaching implications, all of which have been confirmed by experiment. We shall discuss a number of these (time dilation, length contraction, the dependence of mass on velocity, the equivalence of mass and energy, and the impossibility of accelerating a body beyond the speed of light) in the sections that follow.

6.5 TIME DILATION

If an observer measures the time interval between two events which occur <u>at the same point</u> in the frame in which he is at rest (the rest frame) as t_0, then an observer who has a constant velocity v with respect to the first observer will measure the interval as t, where

$$t = \frac{t_0}{\sqrt{1 - v^2/c^2}} \quad ^\star \qquad\qquad [6.1]$$

Equation [6.1] reduces to $t = t_0$ when $v = 0$, but when v has any other value **t is greater than t_0**. (Hence the term 'time dilation'.)

Notes: (i)   Equation [6.1] does not depend on either the position or the direction of motion of the second observer; it depends only on his speed.

(ii)   The events occur at two <u>different</u> points in the rest frame of the second observer.

Suppose that the pilot of an aeroplane shines a lamp for exactly one second as measured by a clock in the plane. The pilot's clock is at rest with respect to the lamp and therefore the time it registers is represented by t_0 in equation [6.1], i.e. $t_0 = 1$ second (exactly). Suppose also that the time for which the lamp is on is measured by a girl standing on the ground directly below. Since the lamp is moving relative to the girl, her clock will register a time represented by t in equation [6.1]. Since t is greater than t_0, she will regard the lamp as being lit for more than one second. (**Note**. This has nothing to do with the fact that the light takes a finite time to reach the girl – we may arrange that the lamp is the same distance from her at the instant it goes off as it was when it came on.)

If the 'experiment' had been performed in reverse, i.e. if the girl on the ground had shone a lamp for exactly one second, the pilot of the plane would have timed it as shining for more than one second (for now t_0 is the time on the girl's clock and t is the pilot's time). Thus **in each case the person who is moving with respect to the lamp regards it as being lit for longer than the person who is at rest with respect to the lamp**.

Whilst the pilot's lamp is on, his clock registers one second and the girl's clock registers more than one second. She therefore regards the pilot's clock as running slow. Whilst the girl's lamp is on, her clock registers one second and the pilot's clock registers more than one second. He therefore regards the girl's clock as running slow. (**Note**. We are not saying that each clock is running slower than the other from any <u>one</u> observer's point of view – this would be impossible. We are saying that <u>one</u> observer regards <u>one</u> clock as running slow and that the <u>other</u> observer regards the <u>other</u> clock as being the slower one.) In general:

> Any observer regards a clock which is moving relative to himself as running slower than a clock which is stationary relative to himself.

The reader may still have some questions

Question   Which clock is <u>really</u> running slower, the girl's or the pilot's?

Answer   The question is meaningless. We have tried to show that time is <u>relative</u> and this question implies that there is some <u>absolute</u> time

---

*See section 6.8 for derivation.

because it does not ask from whose point of view. The most we can say is that the pilot's clock is running slow as far as the girl is concerned and that the girl's clock is the slower one as far as the pilot is concerned.

Question Does a clock change physically as a result of its motion?

Answer No! If it did, it would have to change by different amounts, all at the same time, to accommodate the fact that it is moving at different speeds with respect to different observers.

Question The first 'experiment' shows that the girl regards the pilot's clock as running slower than her own. Why does this not mean that the pilot regards the girl's clock as running faster than his?

Answer Imagine that they can see each other's clocks. The pilot's clock is at rest with respect to the lamp and therefore both the girl and the pilot see it register one second whilst the lamp is on. However, the girl's clock is moving with respect to the lamp and therefore what the pilot sees the girl's clock register whilst his lamp is on is not the same as what the girl sees. The first 'experiment' allows the girl to compare the two clocks but it does not allow the pilot to do so.

6.6 PROPER TIME

The term **proper time** is used to describe the time interval between two events measured in the frame in which they both occur at the same point.

It follows that if a lamp is switched on and then off again, the time for which it is on as measured by a clock which is at rest with respect to the lamp is the proper time. It should be clear that t_0 in equation [6.1] is proper time, and that **the proper time interval between two events is less than any other measurement of the interval**.

QUESTIONS 6A

These questions are based on $t = t_0/\sqrt{1 - v^2/c^2}$. (Take $c = 3.0 \times 10^8 \, \text{m s}^{-1}$ where necessary.)

1. A high-intensity signal light on Earth flashes on for 1.4×10^{-5} s, as measured on Earth, as a spaceship passes overhead moving at $2.7 \times 10^8 \, \text{m s}^{-1}$. For how long is the light on from the point of view of an observer on the spaceship?

2. A spaceship passes the Earth at a speed of $0.8c$ and flashes a signal lamp for $2.0 \, \mu\text{s}$. What is the duration of the signal as measured on Earth?

3. The average lifetime of muons at rest in a laboratory is 2.2×10^{-6} s as measured in the laboratory. The average lifetime of muons moving at high speed in the same laboratory is found to be 6.4×10^{-6} s. At what fraction of c are these muons moving?

4. A spaceship passing the Earth at $1.8 \times 10^8 \, \text{m s}^{-1}$ emits a pulse of light that an observer on Earth measures as lasting for 6.4×10^{-5} s. What is the duration of the pulse as measured on the spaceship?

5. An aeroplane flying at $300 \, \text{m s}^{-1}$ passes over a beacon which flashes on for 6 ms exactly. For how much longer than this would the pilot estimate that the beacon was on. (Unless you have a calculator that can handle 14 significant figures, make use of the approximation $1/\sqrt{1 - v^2/c^2} \approx 1 + \frac{1}{2}v^2/c^2$.)

6.7 THE TWIN PARADOX

We showed in section 6.5 that all observers regard clocks which are moving relative to themselves as running more slowly than their own. It follows that they will also regard people who are moving relative to themselves as ageing more slowly than they themselves do.

Consider two twins, A and B. Suppose that B travels to a distant planet and back, and that A stays behind on Earth. Since each twin is moving relative to the other, it would appear that we could argue that from A's point of view B would age more slowly than A and that from B's point of view A would age more slowly than B. There is clearly a problem with this argument because if A meets up with B on his return, it will immediately be apparent which of the two has aged more. (Assuming, of course, that B has been away long enough for the difference in ageing to be obvious.) This is the so-called 'twin paradox', we seem to be able to argue that each twin will be younger than the other.

The argument fails because B has been at rest in two different inertial frames (one moving away from the Earth and the other moving towards it) whereas A has been at rest with respect to the same inertial frame throughout. The situation does not have the symmetry that this (false) argument assumes. We shall now analyse the situation properly and show that it is B who has aged less than A.

Suppose that B flashes a light in his spaceship as it leaves the Earth, and then again when it reaches the distant planet. The flashes occur at the same point in the frame of the spaceship and therefore the interval between them measured by B is the proper time, t_0. If the speed of the spaceship relative to the Earth is v, then $t = t_0/\sqrt{1 - v^2/c^2}$ where t is the interval between the flashes as measured by A. (Note, this is not the same as the time between the flashes being seen by A.)

If B returns to the Earth in the same time t_0, and flashes a light as he leaves the planet and then again when he reaches the Earth, A will again regard the interval between the flashes as t, where $t = t_0/\sqrt{1 - v^2/c^2}$ and will therefore regard the total journey time as $2t_0/\sqrt{1 - v^2/c^2}$ whereas B regards it as $2t_0$.

If $v = 0.8c$, then $1/\sqrt{1 - v^2/c^2} = 5/3$ and therefore if B regards the journey as taking 6 years, say, A regards it as taking 10 years. Thus, when B returns he is four years younger than his twin.

We have split the journey into two parts because B has been at rest in two different inertial frames and the time dilation formula (equation [6.1]) relates time measured in one inertial frame to that measured in one other, not two. If we had not treated the two parts of the journey separately, we could have had A flashing a light when B left Earth and then again when he returned. This would make the time measured by A the proper time and would produce the incorrect conclusion that A ages less than B. When we split the journey up we need to have A flashing a light when B changes frames, i.e. when B reaches the planet. This is not possible because A does not know when B reaches the planet.

So far we have taken no account of the fact that B has had periods of acceleration and deceleration. If we were to, we would still find that B ages less than A★. In any case, we can assume that the journey is so long that the time spent accelerating and decelerating is negligible in comparison.

★This is in the province of general relativity and we would find that the age difference was even greater than we have calculated here.

Physical Explanation

The reader who is still not convinced that B will age less than A may prefer the following explanation.

Once again, suppose that B travels at a speed of $0.8c$ and, as measured by clocks in the spaceship, takes 3 years to reach the planet and 3 years to return. Suppose that he directs pulses of light towards the Earth at 1 year intervals. (The first of these one year after he leaves and the last of them at the instant he returns.) Bearing in mind that $1/\sqrt{1 - v^2/c^2} = 5/3$ when $v = 0.8c$, A will regard these as intervals of $5/3$ years.

Furthermore, from A's point of view, B will travel $0.8 \times \frac{5}{3} = \frac{4}{3}$ light-years* between pulses. It follows that whilst B is on the outward part of his journey A will receive pulses at intervals of $\frac{5}{3} + \frac{4}{3} = 3$ years, and at intervals of $\frac{5}{3} - \frac{4}{3} = \frac{1}{3}$ year when B is on the return journey. B will send out 3 pulses during each section of the trip and therefore A regards it as lasting $3 \times 3 + 3 \times \frac{1}{3} = 10$ years.

Suppose also that A is sending pulses to the spaceship at intervals of 1 year for 10 years as measured by a clock on Earth. Because of the effects of time dilation B regards these as intervals of $\frac{5}{3}$ years. Whilst he is moving away from the Earth he will receive pulses at 3 year intervals and at intervals of $\frac{1}{3}$ year when he is returning to the Earth. Since he reaches the planet after 3 years of his time, he will receive 1 pulse on the outward journey and 9 on the return journey. B therefore regards the round trip as lasting $1 \times 3 + 9 \times \frac{1}{3} = 6$ years.

This approach demonstrates very clearly that A's point of view is not interchangeable with B's. A receives 3 pulses whilst B is moving away from him and 3 whilst he is returning. If we try to regard B as stationary and A as moving, B receives 1 pulse whilst A is moving away and 9 whilst he is returning. Thus the situation does not have the symmetry that it would have if B kept going in a single direction, i.e. if B were always at rest with respect to the same inertial frame.

6.8 DERIVATION OF $t = t_0/\sqrt{1 - v^2/c^2}$

Consider two inertial frames F and F′, and suppose that F′ is moving in the x direction with a constant velocity v relative to F (Fig. 6.4).

Fig. 6.4
Derivation of
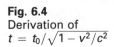

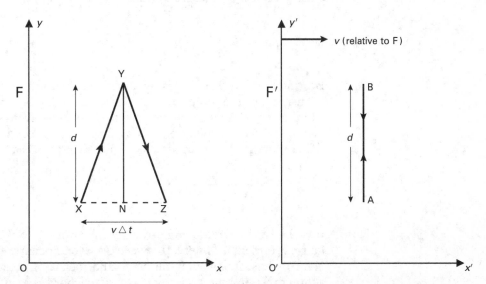

One light-year is a unit of length equal to the distance travelled by light in one year.

Suppose that A and B are two points which are at rest in F' and that a pulse of light travels from A to a mirror at B and then returns to A in a (total) time measured in F' as $\Delta t'$. If $AB = d$, and c is the speed of light then

$$2d = c\Delta t' \tag{6.2}$$

Whilst this is happening, from the point of view of an observer who is at rest in F, frame F' will move to the right by $v\Delta t$, where Δt is the travel time of the pulse as measured in F. (Note, we cannot assume that Δt is the same as $\Delta t'$; indeed the whole point of the present exercise is to show that it is not.) The light path in F will therefore be XYZ where X and Z are the initial and final positions of A, and Y is the position of the mirror when the light reaches it. Since the speed of light is the same for all observers (the second postulate), the speed in F will also be c and therefore

$$XY + YZ = c\Delta t$$

Therefore by Pythagoras' theorem

$$2\sqrt{XN^2 + YN^2} = c\Delta t$$

i.e. $$2\sqrt{\left(\frac{v\Delta t}{2}\right)^2 + d^2} = c\Delta t$$

$$\therefore \quad (v\Delta t)^2 + 4d^2 = (c\Delta t)^2$$

Therefore by equation [6.2]

$$(v\Delta t)^2 + (c\Delta t')^2 = (c\Delta t)^2$$

$$\therefore \quad (c^2 - v^2)(\Delta t)^2 = c^2(\Delta t')^2$$

$$\therefore \quad (1 - v^2/c^2)(\Delta t)^2 = (\Delta t')^2$$

i.e. $$\Delta t = \frac{\Delta t'}{\sqrt{1 - v^2/c^2}} \tag{6.3}$$

Bearing in mind that $\Delta t'$ is the interval as measured in F' between two events (transmission and reception of the light pulse) that occur at the same place in F', and that Δt is the interval between the same two events as measured in a frame moving with a constant velocity v relative to F', equation [6.3] can be written rather more generally, in the notation of section 6.5 as

$$t = \frac{t_0}{\sqrt{1 - v^2/c^2}}$$

Note If we had considered the one-way journey from A to B (X to Y), $\Delta t'$ would no longer represent the interval between two events occurring at the same place in F'. Nevertheless, this would still be a valid means of obtaining equation [6.3] for although B is not the same point as A, its x-coordinate in F' is the same as A's and the (relative) motion is in the x-direction only.

6.9 LENGTH CONTRACTION

We show in section 6.10 that

$$L = L_0\sqrt{1 - v^2/c^2} \qquad [6.4]$$

where L_0 is the length of a rod (say) measured by an observer who is at rest relative to the rod, and L is its length as measured by an observer who is moving with a constant velocity v relative to the rod and parallel to its length.

It follows from equation [6.4] that L is less than L_0 for all non-zero values of v, and therefore:

> All observers regard an object which is moving relative to themselves as being shorter in the direction of motion than do observers who are at rest relative to the object.

Notes (i) L_0 is known as the **proper length**.

(ii) There is no contraction in directions at 90° to the direction of relative motion.

QUESTIONS 6B

The questions that follow are based on equation [6.4].

1. A spaceship of length 60.0 m passes the Earth at a speed of 0.980c. What is the length of the spaceship as measured by an observer on the Earth?

2. An observer measures the length of a metre rule to be 80 cm. At what speed is the rule moving relative to the observer? ($c = 3.0 \times 10^8\,\mathrm{m\,s^{-1}}$.)

EXAMPLE 6.1

A spaceship travels between two beacons, A and B, at a speed of $1.8 \times 10^8\,\mathrm{m\,s^{-1}}$ in a time of 4.0 s as measured in the spaceship. If A and B are at rest relative to the Earth, what is the distance AB from the point of view of **(a)** the pilot of the spaceship, **(b)** an observer at rest on the Earth? ($c = 3.0 \times 10^8\,\mathrm{m\,s^{-1}}$.)

Solution

(a) The pilot moves at $1.8 \times 10^8\,\mathrm{m\,s^{-1}}$ for 4.0 s and therefore regards the distance AB as $1.8 \times 10^8 \times 4.0 = 7.2 \times 10^8\,\mathrm{m}$.

(b) The Earth observer regards the spaceship as travelling for a time t where, by equation [6.1]

$$t = \frac{4.0}{\sqrt{1 - \dfrac{(1.8 \times 10^8)^2}{(3.0 \times 10^8)^2}}} = 5.0\,\mathrm{s}$$

He therefore regards the distance AB as $1.8 \times 10^8 \times 5.0 = 9.0 \times 10^8\,\mathrm{m}$.

This result has been obtained on the basis of time dilation (equation [6.1]) but the reader should realize that it is entirely consistent with length contraction and could have been obtained on the basis of equation [6.4]. We shall now do this.

A and B are at rest with respect to the Earth and are moving with respect to the spaceship. It follows that in the notation of equation [6.4], the pilot's measurement of the distance AB is L and the Earth observer's measurement is L_0.

(a) $\qquad L = 1.8 \times 10^8 \times 4.0 = 7.2 \times 10^8 \text{ m}$

(b) By equation [6.4]

$$7.2 \times 10^8 = L_0 \sqrt{1 - \frac{(1.8 \times 10^8)^2}{(3.0 \times 10^8)^2}}$$

i.e. $\quad L_0 = 9.0 \times 10^8 \text{ m}$

which is the result obtained previously.

Notes (i) What is a time dilation effect to the Earth observer is a length contraction effect to the pilot.

(ii) When solving problems of this type, in which time and distance measurements are 'paired' through the use of 'distance = speed × time', students should be careful to pair only those quantities that are being measured in the same inertial frame, e.g. the pilot's value for time with the pilot's value for distance.

QUESTIONS 6C

1. A muon which has a lifetime of 2.20×10^{-6} s travels through the Earth's atmosphere with a speed of $0.990c$ relative to the Earth. **(a)** What is its lifetime as measuured by an observer on the Earth? **(b)** How far does the Earth observer regard the muon as travelling in this time? **(c)** How far would this appear to be to someone travelling with the muon? ($c = 3.00 \times 10^8 \text{ m s}^{-1}$.)

2. Observers on Earth track an alien spacecraft moving at $0.90c$ for 3.0 years. How far, in light-years, does the craft travel in this time from the point of view of **(a)** the Earth observers, **(b)** the

beings in the spacecraft? **(c)** What time would have elapsed on the spacecraft?

3. The diameter of the Earth's orbit is 3.0×10^{11} m. A spaceship crosses the orbit in 750 s, as measured in the spaceship. What is the speed of the spaceship relative to the Earth? ($c = 3.0 \times 10^8 \text{ m s}^{-1}$.)

4. A spaceship of proper length 70 m passes a space station at a speed of $2.0 \times 10^8 \text{ m s}^{-1}$. How long does it take to pass the space station as measured by **(a)** the pilot of the spaceship, **(b)** an observer on the space station? ($c = 3.0 \times 10^8 \text{ m s}^{-1}$.)

6.10 DERIVATION OF $L = L_0\sqrt{1 - v^2/c^2}$

Consider two inertial frames, F and F', and suppose that F' is moving in the *x*-direction with a constant velocity v relative to F (Fig. 6.5).

Fig. 6.5
Derivation of
$L = L_0\sqrt{1 - v^2/c^2}$

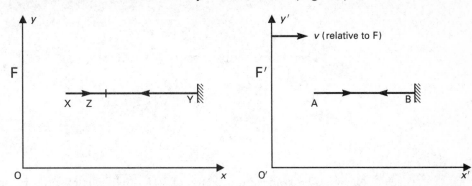

Suppose that A and B are two points which are at rest in F' and that their separation measured in F' is L_0. Suppose that a pulse of light travels from A to a mirror at B and then returns to A in a (total) time $\Delta t'$ as measured in F'. If the speed of light is c, it follows that

$$\Delta t' = 2L_0/c \tag{6.5}$$

Whilst this is happening, from the point of view of an observer who is at rest in F, frame F' will move to the right. The light path in F will be XYZ where X and Z are the initial and final positions of A, and Y is the position of the mirror when the light pulse hits it. If the light takes a time Δt_1 as measured in F to travel from X to Y, then F' will move to the right by $v\Delta t_1$. If the distance AB as measured in F is L, then

$$XY = L + v\Delta t_1$$

Since the speed of light is the same for all observers (the second postulate), the speed in F will also be c and therefore

$$XY = c\Delta t_1$$

$$\therefore \qquad L + v\Delta t_1 = c\Delta t_1$$

i.e. $\qquad \Delta t_1 = \dfrac{L}{c - v} \tag{6.6}$

If the time for the return journey from Y to Z is Δt_2 as measured in F, then

$$YZ = L - v\Delta t_2$$

and

$$YZ = c\Delta t_2$$

$$\therefore \qquad L - v\Delta t_2 = c\Delta t_2$$

i.e. $\qquad \Delta t_2 = \dfrac{L}{c + v} \tag{6.7}$

The total journey time in F is $\Delta t_1 + \Delta t_2$ and is given by equations [6.6] and [6.7] as

$$\Delta t_1 + \Delta t_2 = \frac{L}{c - v} + \frac{L}{c + v}$$

i.e. $\qquad \Delta t_1 + \Delta t_2 = \dfrac{2cL}{c^2 - v^2} \tag{6.8}$

Since $\Delta t'$ is the interval between two events (light leaving A and light returning to A) that occur at the same point in F', and $\Delta t_1 + \Delta t_2$ is the interval between the same two events as measured in F, it follows from the time dilation equation (equation [6.1]) that

$$\Delta t_1 + \Delta t_2 = \frac{\Delta t'}{\sqrt{1 - v^2/c^2}}$$

Therefore by equations [6.5] and [6.8]

$$\frac{2cL}{c^2 - v^2} = \frac{2L_0/c}{\sqrt{1 - v^2/c^2}}$$

$$\therefore \qquad L = \frac{L_0(c^2 - v^2)}{c^2\sqrt{1 - v^2/c^2}} \qquad \text{i.e.} \qquad L = L_0\sqrt{1 - v^2/c^2}$$

Note The contraction factor of $\sqrt{1 - v^2/c^2}$ is exactly the same as that proposed by Lorentz and by Fitzgerald to explain the null result of the Michelson–Morley experiment (section 6.2). However, whereas both Lorentz and Fitzgerald regarded the contraction as an actual (physical) change in length, it emerges from the theory of relativity as a natural consequence of the properties of space and time.

6.11 THE DEPENDENCE OF MASS ON VELOCITY

If the law of conservation of momentum is to be valid in all inertial frames, we have to abandon the idea that the mass of a body is simply a measure of the amount of material it contains and treat it instead as a quantity that increases with velocity.

We shall have cause to use the terms 'rest mass' and 'relativistic mass', and we shall define these before going further.

> **The rest mass** (m_0) of a body is its mass as measured by an observer with respect to whom the body is at rest.
>
> **The relativistic mass** (m) of a body is its mass as measured by an observer with respect to whom it has some velocity other than zero.

It can be shown that

$$m = \frac{m_0}{\sqrt{1 - v^2/c^2}} \tag{6.9}$$

where m_0 is the rest mass of the body and m is its mass as measured by an observer with respect to whom it has velocity v, i.e. its relativistic mass.

Notes (i) Rest mass is sometimes referred to as **proper mass**.

(ii) The validity of equation [6.9] has been confirmed by experiment (from studies of high-energy electrons, for example) on many occasions.

(iii) $\qquad \text{Momentum } (p) = mv = \dfrac{m_0 v}{\sqrt{1 - v^2/c^2}}$

The Impossibility of $v > c$

The denominator of equation [6.9] becomes smaller as v increases, i.e. **relativistic mass increases with velocity**. (See Fig. 6.6.) Unless $m_0 = 0$, m approaches infinity as v approaches c. This implies that **a body with non-zero rest mass★ cannot be accelerated up to the speed of light**, for if it were to reach the speed of light, it would have infinite mass and this would be absurd. Thus bodies of non-zero rest mass (with the possible exception of tachyons) must always travel at speeds less than the speed of light.

Fig. 6.6
Relativistic mass as a function of speed

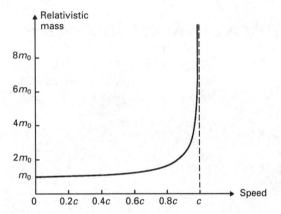

The reader may feel that there are two possible objections to this statement.

1. Suppose that two particles are moving in opposite directions, each at a speed of $2c/3$ with respect to some observer. Newtonian mechanics gives their relative speed as $4c/3$, which is clearly greater than the speed of light. According to relativity theory, though, the relative speed v of two particles moving in opposite directions with speeds u_1 and u_2 is given by

$$v = \frac{u_1 + u_2}{1 + \dfrac{u_1\,u_2}{c^2}}$$

   which gives a relative speed of $12c/13$ in this case and this is less than the speed of light. The reader should confirm that even when $u_1 = u_2 = c$, the relative speed is c!

2. Suppose that a powerful laser beam is directed towards the Moon so that a spot of light falls on its surface. If the laser were then turned sideways at the relatively moderate rate of 180 degrees per second, the spot would move across the Moon at a speed in excess of the speed of light. There is nothing in the theory of relativity to preclude this; neither matter, nor energy, nor information has moved across the Moon.

★Particles with zero rest mass (e.g. photons and neutrinos) always travel at the speed of light (see section 6.14). It has been suggested (by E.C.G. Sudarshan) that there might be a group of particles (called **tachyons**) for which $v > c$ at the instant they are created and which would always travel faster than light and which would speed up as they lost energy! They have never been detected.

QUESTIONS 6D

1. What is the relativistic mass of a proton moving at: **(a)** $0.90c$, **(b)** $0.94c$, **(c)** $0.98c$? (Rest mass of proton $= 1.67 \times 10^{-27}$ kg, $c = 3.00 \times 10^8$ m s^{-1}.)

2. At what speed is a particle moving if the ratio of its relativsitic mass to its rest mass is: **(a)** 2, **(b)** 10? ($c = 3.00 \times 10^8$ m s^{-1}.)

3. By what factor does the momentum of a particle increase when its speed is doubled: **(a)** from $0.40c$ to $0.80c$, **(b)** from $0.49c$ to $0.98c$?

6.12 DERIVATION OF $E = mc^2$

A mass m is equivalent to an amount of energy E where $E = mc^2$. In order to derive this result we shall consider a **thought experiment**, i.e. an experiment which can be carried out in principle only.

Fig. 6.7
Derivation of $E = mc^2$

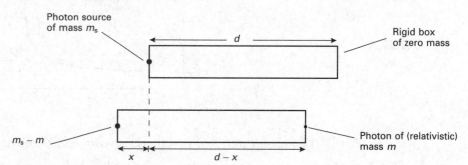

Consider a rigid box of length d and of zero mass, with a 'photon source' (e.g. a very low power lamp) attached to one end (Fig. 6.7). Suppose that the source emits a photon of energy E and momentum E/c which moves to the other end of the box where it is absorbed. (It follows from Maxwell's equations that a pulse of light of energy E has momentum E/c and therefore a photon of energy E has momentum E/c.) The box recoils when the photon is emitted and is subsequently brought to rest when the photon is absorbed. If the box recoils through a distance x, the photon will travel a distance $(d - x)$ in a time $(d - x)/c$ before being absorbed. If the recoil velocity of the box is v, the box will move a distance $v(d - x)/c$ in this time and therefore

$$x = v(d - x)/c \qquad [6.10]$$

If the original mass of the source is m_s, and the photon has (relativistic) mass m, then by the principle of conservation of momentum

$$(m_s - m)v = \frac{E}{c}$$

Therefore by equation [6.10]

$$x = \frac{E(d - x)}{(m_s - m)c^2} \qquad [6.11]$$

No external force has acted on the system and therefore its centre of mass cannot have moved. It follows that it must still be at the original position of the photon source and therefore (since there can be no resultant turning moment about the centre of mass)

$$(m_s - m)x = m(d - x)$$

Substituting for x from equation [6.11] gives

$$\frac{(m_s - m) E (d - x)}{(m_s - m) c^2} = m (d - x)$$

i.e. $E = mc^2$

Although we have derived this result by considering a photon, there is no reason to suppose that it is not completely general, for all energy becomes electromagnetic radiation eventually. The result has, in any case, been confirmed by experiment – for example, by comparing the decrease in mass that occurs in fission and fusion reactions with the energy released. Confirmation also comes from the decay of (so-called) fundamental particles such as the π^0 meson. This decays to produce two γ-ray photons – measurements show that the energy of the photons is equal to $m_\pi c^2$ where m_π is the mass of the meson.

6.13 THE EQUIVALENCE OF MASS AND ENERGY

We have shown in section 6.12 that a mass m is equivalent to an amount of energy E where

$$E = mc^2 \qquad [6.12]$$

The energy of a body which is at rest is the energy associated with its rest mass and is called the **rest energy** of the body. The rest energy, E_0, of a body of rest mass m_0 is given by equation [6.12] as

$$E_0 = m_0 c^2$$

If this same body is moving with speed v, its relativistic mass is m, where

$$m = m_0 (1 - v^2/c^2)^{-1/2} \qquad \text{(equation [6.9])}$$

and its **total energy**, E, is given by

$$E = mc^2 = m_0 c^2 (1 - v^2/c^2)^{-1/2}$$

The relativistic kinetic energy of a body is the energy associated with its motion and is defined by

Total energy $=$ Rest energy $+$ Relativistic KE

i.e. $E = E_0 + K \qquad [6.13]$

where K is the relativistic kinetic energy.

Rearranging equation [6.13] gives

$$K = E - E_0$$

$$= m_0 c^2 (1 - v^2/c^2)^{-1/2} - m_0 c^2$$

$$= m_0 c^2 \left(1 + \frac{1}{2} \frac{v^2}{c^2} + \frac{3}{8} \frac{v^4}{c^4} + \dots \right) - m_0 c^2$$

i.e. $K = m_0 c^2 \left(\dfrac{1}{2} \dfrac{v^2}{c^2} + \dfrac{3}{8} \dfrac{v^4}{c^4} + \dots \right)$

When $v \ll c$, the only term of any significance on the right-hand side of this expression is the first one, and therefore to a good approximation

$$K = \frac{1}{2}m_0 v^2$$

which is the classical (Newtonian) expression for kinetic energy.

The realization that mass and energy are equivalent and that a body has energy (its rest energy) simply because it has mass means that the principles of conservation of mass and of conservation of energy (of pre-relativity physics) are no longer valid in all circumstances. In pre-relativity physics 'mass' meant what is now called rest mass and 'energy' meant energy other than rest energy. There are situations in which neither of these quantities is conserved and we must think instead in terms of **the principle of conservation of mass and energy**, for it is the total energy, E, that is conserved. (**Note**. This is equivalent to saying that relativistic mass is conserved.)

EXAMPLE 6.2

Two identical particles whose rest masses are m_0 are moving in opposite directions, each with velocity v. They collide inelastically and coalesce to produce a single particle of rest mass M_0. Find M_0 in terms of m_0, v and c.

Solution

The two original particles have equal and opposite momenta and therefore it follows from the principle of conservation of momentum that the composite particle will be at rest when it is formed. Its total energy will therefore be its rest energy $M_0 c^2$.

Total energy is conserved, and therefore

$$\frac{m_0 c^2}{\sqrt{1 - v^2/c^2}} + \frac{m_0 c^2}{\sqrt{1 - v^2/c^2}} = M_0 c^2$$

i.e. $$M_0 = \frac{2m_0}{\sqrt{1 - v^2/c^2}}$$

Thus $M_0 > 2m_0$, i.e. the total rest mass has increased. The reader might like to confirm that the increase in rest mass is equal to the decrease in relativistic KE divided by c^2.

EXAMPLE 6.3

Calculate the speed of an electron which has been accelerated from rest through a PD of 2.0×10^6 V.

(Rest mass of electron = 9.1×10^{-31} kg, $c = 3.0 \times 10^8$ m s^{-1}, $e = 1.6 \times 10^{-19}$ C.)

Solution

Since $E = mc^2$

$$\begin{pmatrix} \text{Increase} \\ \text{in energy} \end{pmatrix} = \begin{pmatrix} \text{Increase in} \\ \text{relativistic mass} \end{pmatrix} c^2$$

Therefore in the usual notation

$$eV = (m - m_0) c^2 \; \star$$

$$\therefore \quad eV = m_0 \left(\frac{1}{\sqrt{1 - v^2/c^2}} - 1 \right) c^2$$

$$\therefore \quad \frac{1}{\sqrt{1 - v^2/c^2}} = \frac{eV}{m_0 c^2} + 1$$

$$= \frac{(1.6 \times 10^{-19}) (2.0 \times 10^6)}{(9.1 \times 10^{-31}) (3.0 \times 10^8)^2} + 1 = 4.91$$

$$\therefore \quad 1 - v^2/c^2 = \frac{1}{4.91^2} = 4.15 \times 10^{-2}$$

$$\therefore \quad \frac{v^2}{c^2} = 0.959$$

$$\therefore \quad \frac{v}{c} = 0.979$$

i.e. $\quad v = 0.979 \times 3.0 \times 10^8 = 2.9 \times 10^8 \, \text{m s}^{-1}$

QUESTIONS 6E

1. Find: **(a)** the relativistic mass, **(b)** the relativistic momentum, **(c)** the total energy, **(d)** the rest energy and **(e)** the relativistic KE of an electron which has a velocity of $2.8 \times 10^8 \, \text{m s}^{-1}$. (Rest mass of electron $= 9.1 \times 10^{-31}$ kg, $c = 3.0 \times 10^8 \, \text{m s}^{-1}$.)

2. Calculate the energy released when a π^0 meson decays to produce electromagnetic radiation. (Rest mass of π^0 meson $= 2.4 \times 10^{-28}$ kg, $c = 3.00 \times 10^8 \, \text{m s}^{-1}$.)

3. An electron is accelerated from rest through a PD of 1.00×10^5 V. Calculate its speed on the basis of **(a)** relativistic mechanics, **(b)** Newtonian mechanics. (Rest mass of electron $= 9.11 \times 10^{-31}$ kg, $c = 3.00 \times 10^8 \, \text{m s}^{-1}$, $e = 1.60 \times 10^{-19}$ C.)

4. Through what PD must an electron be accelerated to increase its speed to $0.960c$ from rest? (Rest mass of electron $= 9.11 \times 10^{-31}$ kg, $c = 3.00 \times 10^8 \, \text{m s}^{-1}$, $e = 1.60 \times 10^{-19}$ C.)

5. Through what PD must an electron be accelerated to increase its speed from $0.950c$ to $0.990c$? (Rest mass of electron $= 9.11 \times 10^{-31}$ kg, $c = 3.00 \times 10^8 \, \text{m s}^{-1}$, $e = 1.60 \times 10^{-19}$ C.)

6. A particle of rest mass m_0 moving at a speed of $0.90c$ collides inelastically with an identical particle which is at rest. They join to produce a single particle of rest mass M_0 moving with speed v. Use the principles of conservation of momentum and of (total) energy to find: **(a)** v in terms of c, and **(b)** M_0 in terms of m_0.

7. What is: **(a)** the speed, **(b)** the relativistic KE of a proton whose momentum is $4.00 \times 10^{-19} \, \text{kg m s}^{-1}$? (Rest mass of proton $= 1.67 \times 10^{-27}$ kg, $c = 3.00 \times 10^8 \, \text{m s}^{-1}$.)

$\star$The electron is initially at rest and therefore its initial <u>relativistic</u> mass is its rest mass, m_0.

6.14 THE ENERGY–MOMENTUM EQUATION

A particle with relativistic mass m has total energy E where

$$E = mc^2$$

$$\therefore \quad E^2 = m^2 c^4$$

If the particle has rest mass m_0 and velocity v

$$E^2 = \frac{m_0{}^2}{1 - v^2/c^2} c^4$$

$$= m_0{}^2 c^4 \left[\frac{1}{1 - v^2/c^2} - 1 \right] + m_0{}^2 c^4$$

$$= m_0{}^2 c^4 \left[\frac{v^2/c^2}{1 - v^2/c^2} \right] + m_0{}^2 c^4$$

$$= \frac{m_0{}^2 v^2}{1 - v^2/c^2} c^2 + m_0{}^2 c^4$$

i.e. $$E^2 = p^2 c^2 + m_0{}^2 c^4 \qquad\qquad\qquad\qquad [6.14]$$

where p is the (relativistic) momentum of the particle.

6.15 PARTICLES OF ZERO REST MASS

Particles of zero rest mass (e.g. photons and neutrinos) always travel at the speed of light (in vacuum). Even though their rest mass is zero, they have both energy and momentum.

For a particle of total energy E, momentum p, and rest mass m_0

$$E^2 = p^2 c^2 + m_0{}^2 c^4 \qquad \text{(equation [6.14])}$$

Therefore for a particle of zero rest mass

$$E^2 = p^2 c^2$$

i.e. $$E = pc \qquad\qquad\qquad\qquad\qquad\qquad\qquad [6.15]$$

If we regard the particle as having relativistic mass m and speed v, then $E = mc^2$ and $p = mv$, in which case by equation [6.15]

$$mc^2 = mvc$$

i.e. $$v = c$$

confirming that particles of zero rest mass travel at the speed of light. Summarizing

$$\text{When } m_0 = 0 \begin{cases} E = pc \\ v = c \\ m = \dfrac{E}{c^2} = \dfrac{p}{c} \end{cases}$$

6.16 EVIDENCE FOR TIME DILATION FROM MUON DECAY

Cosmic ray bombardment of atoms in the upper atmosphere can produce muons – unstable particles which then travel towards the surface of the Earth at speeds close to the velocity of light. In an experiment performed in 1963, Frisch and Smith obtained data on these fast-moving muons which provided convincing evidence for the validity of the time dilation equation (equation [6.1]).

Fig. 6.8
Experiment to confirm
time dilation

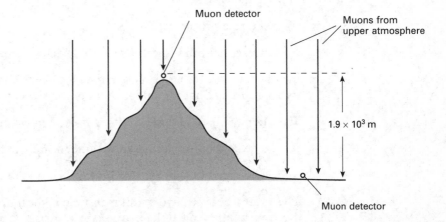

Frisch and Smith compared the rate at which muons moving at $0.994c$ reached a detector on the top of Mt. Washington (USA) with the rate at which muons of the same speed reached a detector 1.9×10^3 m below (Fig. 6.8). As measured by an observer on Earth, this journey takes a time t, where

$$t = \frac{1.9 \times 10^3}{0.994 \times 3.0 \times 10^8} = 6.37 \, \mu s$$

The half-life of muons as measured by an observer relative to whom they are at rest is $1.53 \, \mu s$. Frisch and Smith found that the fraction remaining after travelling 1.9×10^3 m was 0.732, which implies that an observer moving along with the muons would regard the journey as taking a time t_0 given by the exponential law of radioactive decay as

$$0.732 = e^{-(0.6931 \, t_0 / 1.53 \times 10^{-6})}$$

i.e. $t_0 = 0.689 \, \mu s$

The time dilation factor is therefore $6.37/0.689 = 9.2$, which compares well with that calculated on the basis of equation [6.1], namely $1/\sqrt{1 - 0.994^2} = 9.1$.

CONSOLIDATION

The Ether

The hypothetical medium that was once considered necessary to account for the fact that light can travel through a vacuum.

It was regarded as a perfect frame of reference relative to which all motion could be measured and which therefore gave meaning to the concept of absolute motion.

Light was supposed to travel at a fixed speed (c) with respect to the ether.

The Michelson–Morley Experiment

The aim of the experiment was to measure the speed of the Earth with respect to the ether and therefore

(i)   to confirm the existence of the ether, and

(ii)   to determine the <u>absolute</u> speed of the Earth.

The null result (i.e. the failure to detect any motion of the Earth through the ether) implies that either

(i)   the ether does not exist, or

(ii)   it does exist but is impossible to detect.

In either case we have to abandon the idea of absolute motion and have to accept that the speed of light is independent of the velocity of the observer.

The theory of special relativity is based on two postulates

1.   The laws of physics have the same form in all inertial frames of reference.

2.   The speed of light (in vacuum) is the same for all inertial observers.

The interval between two events measured in the frame in which they both occur at the <u>same point</u> is called the **proper time** (t_0). The interval (t) between the same two events as measured in a frame which has a constant velocity v with respect to the first frame is given by

$$t = \frac{t_0}{\sqrt{1 - v^2/c^2}} \qquad \textbf{(Time dilation)}$$

If L_0 is the length of a rod as measured by an observer who is at rest relative to the rod, then an observer moving at a constant velocity v relative to the rod regards its length as L where

$$L = L_0\sqrt{1 - v^2/c^2} \qquad \textbf{(Length contraction)}$$

The rest mass (m_0) of a body is its mass as measured by an observer with respect to whom the body is at rest.

The relativistic mass (m) of a body is its mass as measured by an observer with respect to whom it has some velocity other than zero.

$$m = \frac{m_0}{\sqrt{1 - v^2/c^2}}$$

$$p = mv = \frac{m_0 v}{\sqrt{1 - v^2/c^2}}$$

$$E = mc^2 = \frac{m_0 c^2}{\sqrt{1 - v^2/c^2}}$$

$$E^2 = p^2 c^2 + m_0^2 c^4$$

Total energy (E) = Rest energy (E_0) + Relativistic kinetic energy (K)

$$v = \frac{u_1 + u_2}{1 + \dfrac{u_1 u_2}{c^2}}$$

QUESTIONS ON CHAPTER 6

1. **(a)** Why did scientists once believe in the existence of the ether?
   (b) Why was the concept eventually abandoned?

2. Give a brief account of the Michelson–Morley experiment. Explain how the outcome of the experiment implied that the concept of absolute motion is meaningless.

3. An atomic clock moves at $300\,\mathrm{m\,s^{-1}}$ for 24.0 hours as measured by an identical stationary clock. How many nanoseconds does the moving clock 'lose' compared with the stationary clock? (Make use of the approximation $(1-x)^{-1/2} \approx 1 + \frac{1}{2}x$.) (Speed of light $= 3.00 \times 10^{8}\,\mathrm{m\,s^{-1}}$.)

4. A star is 8 light-years away from the Earth. A rocket leaves Earth and travels to the star in 6 years as measured by a clock in the rocket.
   (a) In terms of the speed of light, c, what is the speed of the rocket relative to the Earth?
   (b) How long does the journey take as measured by a clock on the Earth?

5. The mean lifetime of muons at rest is $2.2\,\mu s$. An observer notes that they travel an average of $2000\,\mathrm{m}$ before decaying. In terms of the speed of light, c, what is the speed of the muons relative to the observer?

6. **(a)** State the two postulates of Einstein's special theory of relativity.
   (b) Muons are created in the upper atmosphere by cosmic rays. They are negatively charged particles with a mass two hundred times that of an electron and a charge of the same size and sign as an electron. They are very short-lived, decaying into an electron and two neutrinos. The graph illustrates the short-lived nature of stationary muons: it shows the number, N, of muons surviving against time, t.

   For every 1000 muons detected at a height of $2000\,\mathrm{m}$, 700 are detected at sea level.
   (i) Use the graph to estimate how long it would take for 1000 stationary muons to decay to 700.
   (ii) How far would a light photon moving through the atmosphere travel in this time.

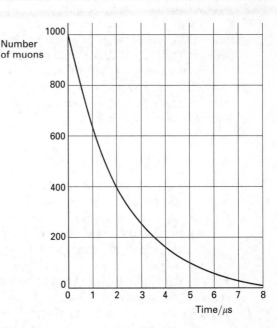

(iii) Muons produced by cosmic radiation travel at a speed of 99.8% of the speed of light. Use the theory of special relativity to explain why such a high percentage of the muons produced by cosmic radiation reach sea level. [J (specimen), '96]

7. Calculate the percentage increase in mass when a particle is accelerated from rest to 1% of the speed of light.

8. A metal cube has a density of $9.0 \times 10^{3}\,\mathrm{kg\,m^{-3}}$ when at rest in a laboratory. If the cube is caused to move perpendicular to two of its faces at a speed of $2.4 \times 10^{8}\,\mathrm{m\,s^{-1}}$ relative to an observer in the laboratory, what is its density as measured by this observer?
   (Speed of light $= 3.0 \times 10^{8}\,\mathrm{m\,s^{-1}}$.)

9. A particle is travelling at 60% of the speed of light. What is the ratio of **(a)** its rest mass to its relativistic mass, **(b)** its kinetic energy calculated on the basis of Newtonian mechanics to that on the basis of relativistic mechanics?

10. **(a)** A high-energy gamma-ray photon can spontaneously create an electron and a positron:
$$\gamma \rightarrow e^{-} + e^{+}$$

Assuming that the rest mass of an electron and that of a positron are each 9.1×10^{-31} kg, that $c = 3.0 \times 10^8$ m s^{-1} and that the Planck constant, h, is 6.6×10^{-34} J s, calculate the *maximum possible* wavelength of the photon.

(b) According to special relativity theory the inertial mass, m, of an electron moving with speed v is given by

$$m = m_0 \left(1 - \frac{v^2}{c^2}\right)^{-1/2}$$

where m_0 is the rest mass of the electron and c is the speed of light *in vacuo*.

(i) Use the equation to explain what happens to the mass m if the electron is accelerated to speeds very close to that of light.

How does the theory forbid electrons from travelling at speeds greater than c?

(ii) Describe one additional relativistic effect exhibited by high speed electrons.   [L, '91]

11. By considering a photon in a box, derive the expression $E = mc^2$.

12. Can a particle with a rest mass m_0 have a momentum of $m_0 c$, where c is the speed of light? Explain your answer.

13. An electron and a positron, each with negligible kinetic energy, annihilate each other and create two identical photons. Calculate **(a)** the energy released by the annihilation, **(b)** the frequency of the photons. (Mass of electron = mass of positron = 9.11×10^{-31} kg, speed of light = 3.00×10^8 m s^{-1}, Planck's constant = 6.63×10^{-34} J s.)

14. Explain how the principle of conservation of energy has had to be modified in the light of the theory of special relativity.

15. What percentage error is introduced when the kinetic energy of a particle moving at 10% of the speed of light is calculated on the basis of Newtonian mechanics rather than relativistic mechanics?

16. A particle of rest mass m_0 has a kinetic energy of $\frac{1}{2} m_0 c^2$. What is the speed of the particle in terms of the speed of light, c?

17. On the basis of 'force = rate of change of (relativistic) momentum', show that a particle of rest mass m_0 acted on by a force F acquires an acceleration dv/dt given by

$$F = m_0 \gamma^3 dv/dt \quad \text{where} \quad \gamma = (1 - v^2/c^2)^{-1/2}$$

18. An electron has a kinetic energy of 3.3×10^{-13} J. At what fraction of the speed of light is the electron moving?
(Mass of electron = 9.1×10^{-31} kg, speed of light = 3.0×10^8 m s^{-1}.)

19. What is **(a)** the speed, **(b)** the momentum of an electron that has been accelerated from rest through a potential difference of 20 000 V?
(Charge on electron = 1.6×10^{-19} C, mass of electron = 9.11×10^{-31} kg, speed of light = 3.00×10^8 m s^{-1}.)

20. The Stanford linear accelerator is 3.0 km long and accelerates electrons to an energy of 20 GeV. What is the length of the accelerator in the rest frame of the electrons?
(Charge on electron = 1.6×10^{-19} C, mass of electron = 9.1×10^{-31} kg, speed of light = 3.0×10^8 m s^{-1}.)

7

LOW TEMPERATURE PHYSICS

The reader should be familiar with the gas laws (Boyle's law, Charles' law and the pressure law) and with the concept of an ideal gas before continuing.

7.1 ABSOLUTE ZERO

Suppose that the volume of a gas is measured as a function of temperature at constant pressure. For moderate pressures and at temperatures well above that at which the gas would liquefy, a graph of volume against temperature is, to a good approximation, a straight line. Repeating the experiment at a lower constant pressure gives an even straighter line, and it can be shown that if the experiment could be carried out at zero pressure, the graph would be exactly linear. Furthermore, if such as graph were extrapolated to zero volume, it would cut the temperature axis at $-273.15\,°C$ and would do so no matter which particular gas were involved (Fig. 7.1). Since a gas at zero pressure behaves as if it is an ideal gas, we can conclude that **the volume of an ideal gas is zero at $-273.15\,°C$**. A lower temperature would imply a negative volume. Since such a concept is meaningless, this suggests that it is impossible to attain temperatures below **$-273.15\,°C$** and therefore that this temperature can meaningfully be regarded as the zero of temperature. It is known as **absolute zero** and is allotted the value of zero kelvin.

Fig. 7.1
Graph of volume against Celsius temperature for an ideal gas

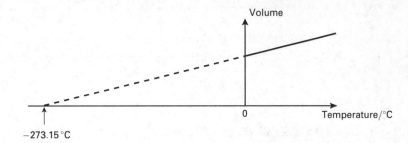

Notes

(i) Although temperatures as low as 10^{-5} K have been attained, **the third law of thermodynamics** states that it will never be possible to reach absolute zero itself. However, the third law is not unanimously accepted and needs to be treated with some caution – it may yet prove possible to reach absolute zero and work is currently underway to do just this.

(ii) Although the molecules of an ideal gas would have zero kinetic energy at zero kelvin, those of a real substance would not. According to quantum theory, all the particles in any given system would be in the lowest available energy states but their total energy would not be zero; they would possess what is known as **zero-point energy**.

Fig. 7.2
Apparatus to estimate
absolute zero

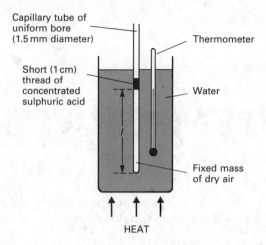

A reasonable estimate of absolute zero can be made using the apparatus in Fig. 7.2, in which a column of air is trapped inside a capillary tube by a short thread of concentrated sulphuric acid. The reason for using the acid (rather than mercury, say) is that it absorbs any water that might be in the air and so allows meaningful results to be obtained. The tube has a <u>uniform</u> bore and therefore the volume of the trapped air is proportional to the length of the air column. The water is heated and the length, l, of the air column is measured for a number of different temperatures, θ. The water should be heated slowly, and stirred before each reading, to allow the air to reach the temperature of the water. The pressure of the air throughout the experiment is <u>constant</u> (equal to atmospheric pressure plus the pressure exerted by the acid thread).

A graph of l against <u>Celsius</u> temperature, θ, is plotted. By extrapolating the graph to zero volume an estimate of absolute zero is obtained (Fig. 7.3).

Fig. 7.3
Graph to estimate
absolute zero

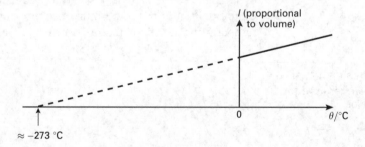

Note   Small values of l should be avoided, otherwise the rounded end of the capillary tube introduces a significant error into the assumption that the volume of trapped air is proportional to l.

7.2 REVERSIBLE PROCESSES

If, at every stage, a process can be made to go in the reverse direction by an infinitesimal change in the conditions which are causing it to take place, it is said to be a reversible process.

It follows that when the state of a system is changed reversibly:

(i)   the system is in **thermodynamic equilibrium** (i.e. all parts of the system are at the same temperature and pressure) at every instant, and

(ii)   at the completion of the process the system could be returned to its initial state by passing through the intermediate states in reverse order, and without there being any net change in the rest of the Universe.

In practice, it is impossible to produce a perfect reversible change. However, processes which take place very slowly and which do not involve friction are often good approximations to reversible changes. The slow compression of a gas by the movement of a light, frictionless piston in a non-conducting cylinder is an example of an approximately reversible process, because a slight decrease in the force on the piston would allow the gas to expand and no energy will have been dissipated as heat. Other examples include the changes of pressure, volume and temperature which are associated with the passage of a sound wave through air, and the movement of a pendulum about a frictionless support in a vacuum.

7.3 EXTERNAL WORK DONE BY AN EXPANDING GAS

Consider a gas enclosed in a cylinder by a frictionless piston of cross-sectional area A (Fig. 7.4). Suppose that the piston is in equilibrium under the action of the force pA exerted by the gas and an external force F. Suppose now, that the gas expands

Fig. 7.4
Gas expanding in a cylinder

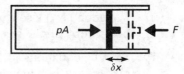

and moves the piston outwards through a distance δx, where δx is so small that p can be considered to be constant. The external work done δW by the expansion is given by

$$\delta W = pA\,\delta x$$

i.e. $\delta W = p\,\delta V$

where δV is the small increase in volume of the gas. The total work done W by the gas if its volume changes by a finite amount from V_1 to V_2 is therefore given by

$$W = \int_{V_1}^{V_2} p\,\mathrm{d}V \qquad\qquad [7.1]$$

Equation [7.1] holds no matter what the relationship between p and V. For the particular case of an **isobaric process** (i.e. one in which p is constant)

$$W = \int_{V_1}^{V_2} p\,\mathrm{d}V = p\int_{V_1}^{V_2} \mathrm{d}V$$

i.e. $W = p(V_2 - V_1)$

Putting $V_2 - V_1 = \Delta V$ gives

$$W = p\Delta V \qquad \text{(at constant pressure)} \qquad [7.2]$$

where W is the work done by a gas expanding by an amount ΔV at a constant pressure p.

In the general case, if a plot of p against V is available (known as an **indicator diagram**), the work done can be obtained graphically. Suppose that the pressure of a gas varies with volume as shown in Fig. 7.5. The work done W by the gas as its volume changes from V_1 to V_2 is given by

$$W = \int_{V_1}^{V_2} p\,dV = \text{Area of shaded region}$$

Fig. 7.5
Indicator diagram for a gas

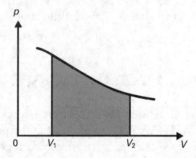

Notes (i) Equation [7.1] also applies when a gas is compressed, in which case work is being done <u>on</u> the gas.

(ii) Strictly, equations [7.1] and [7.2] can be applied only if the change takes place reversibly – if it does not, the values of pressure and temperature at any instant will be different in different regions of the gas.

(iii) Equation [7.1] also applies to solids and liquids. In these cases, though, the increases in volume are small and therefore the amounts of external work done are small compared with increases in internal energy.

7.4 THE FIRST LAW OF THERMODYNAMICS

Thermodynamics is the study of the relationship between heat and other forms of energy. When the principle of conservation of energy is stated with reference to heat and work it is known as the **first law of thermodynamics**.

> The heat energy (ΔQ) supplied to a system is equal to the increase in the internal energy (ΔU) of the system plus the work done (ΔW) by the system on its surroundings.

i.e. $$\Delta Q = \Delta U + \Delta W \qquad [7.3]$$

The internal energy of a system is the sum of the kinetic and potential energies of the <u>molecules of the system</u>. It follows from equation [7.3] that it may be increased by:

(i)   putting heat energy into the system, and/or

(ii)   doing work on the system.

When the internal energy of a system changes the change depends only on the initial and final states of the system, and not on how the change was brought about. This is equivalent to saying that the internal energy of a system depends only on the state that it is in, and not on how it reached that state. (**Note**. A system is said to have changed 'state' if some observable property of the system, e.g. its temperature, pressure, or phase, has changed.)

An **isolated system** is one which is cut off from any form of external influence. In particular, no work can be done on it or by it (i.e. $\Delta W = 0$), and no heat can enter it or leave it (i.e. $\Delta Q = 0$). It follows from equation [7.3] that $\Delta U = 0$, and therefore that **the internal energy of an isolated system is constant**.

When a system undergoes an **adiabatic process** (see section 7.12) $\Delta Q = 0$, and equation [7.3] reduces to $\Delta U = -\Delta W$. Bearing in mind that ΔW represents work done by the system, $(-\Delta W)$ represents work done on the system. Thus, **when a system undergoes an adiabatic process the increase in internal energy of the system is equal to the work done on it**.

7.5 LATENT HEAT

It is necessary to supply energy (heat) to a solid in order to melt it, even if the solid is already at its melting point. This energy is called **latent heat**. It is distinct from any heat that might have been used to bring the solid up to its melting point in the first place, and from that which might be used to raise the temperature of the liquid once the solid has melted.

The energy is used to provide the increased molecular potential energy of the liquid phase and, when the phase change results in expansion, to do external work in pushing back the atmosphere. The energy used to do external work is usually much less than that used to increase the potential energy of the molecules, and in the case of ice, which contracts on melting, is negative.

The conversion of a liquid to a vapour (vaporization) and the direct conversion of a solid to a vapour (sublimation) also require latent heat to be supplied. These two processes usually involve large changes in volume, and the proportion of the latent heat which is used to do external work is greater than in melting.

In terms of the first law of thermodynamics (section 7.4) melting (i.e. fusion), vaporization and sublimation are represented by

$$L = \Delta U + \Delta W$$

where

L = the latent heat supplied in order to cause the phase change

ΔU = the increase in internal potential energy which accompanies the phase change. (There is no change in temperature and therefore no change in kinetic energy.)

ΔW = the external work done as a result of the phase change. This term is positive for expansion and negative for contraction.

> **The specific latent heat** (l) of fusion (or vaporization or sublimation) of a substance is defined as the energy required to cause unit mass of the substance to change from solid to liquid (or liquid to vapour, or solid to vapour) without temperature change. (Unit $= $ J kg^{-1}.)

Note   The value of l depends on the temperature (and therefore the pressure) at which it is measured.

It follows that the heat, ΔQ, which must be added to change the phase of a mass, m, of substance is given by

$$\Delta Q = ml$$

where l is the specific latent heat of fusion, vaporization or sublimation according to the particular phase change which is taking place. For the reverse processes (liquid to solid, vapour to liquid, and vapour to solid) ΔQ represents the amount of heat that must be <u>removed</u> from the substance.

EXAMPLE 7.1

A calorimeter with a heat capacity of 80 J °C^{-1} contains 50 g of water at 40 °C. What mass of ice at 0 °C needs to be added in order to reduce the temperature to 10 °C? Assume no heat is lost to the surroundings. (Specific heat capacity of water $= 4.2 \times 10^3$ J kg^{-1} °C^{-1}, specific latent heat of ice $= 3.4 \times 10^5$ J kg^{-1}.)

Solution

Heat lost by calorimeter cooling to 10 °C

$$= 80(40 - 10) = 2400 \text{ J}$$

Heat lost by water cooling to 10 °C

$$= 50 \times 10^{-3} \times 4.2 \times 10^3 (40 - 10) = 6300 \text{ J}$$

$\therefore$ Total heat lost $= 2400 + 6300 = 8700$ J

Let mass of ice $= m$

Heat used to melt ice at 0 °C

$$= m \times 3.4 \times 10^5 = 3.4 \times 10^5 m$$

Heat used to increase temperature of melted ice to 10 °C

$$= m \times 4.2 \times 10^3 (10 - 0) = 4.2 \times 10^4 m$$

$\therefore$ Total heat used $= 3.4 \times 10^5 m + 4.2 \times 10^4 m = 3.82 \times 10^5 m$

Since no heat is lost to the surroundings,

$$3.82 \times 10^5 m = 8700$$

$\therefore$ $m = 0.0228$ kg

i.e.   Mass of ice required $= 23$ g

QUESTIONS 7A

1. Calculate the heat required to melt 200 g of ice at 0 °C.
   (Specific latent heat of ice $= 3.4 \times 10^5 \, \text{J kg}^{-1}$.)

2. Calculate the heat required to turn 500 g of ice at 0 °C into water at 100 °C.
   (Specific latent heat of ice $= 3.4 \times 10^5 \, \text{J kg}^{-1}$, specific heat capacity of water $= 4.2 \times 10^3 \, \text{J kg}^{-1} \, °\text{C}^{-1}$.)

3. Calculate the heat given out when 600 g of steam at 100 °C condenses to water at 20 °C.
   (Specific latent heat of steam $= 2.26 \times 10^6 \, \text{J kg}^{-1}$, specific heat capacity of water $= 4.2 \times 10^3 \, \text{J kg}^{-1} \, °\text{C}^{-1}$.)

7.6 REAL GASES

When real gases are subjected to pressures which are greater than a few atmospheres and/or when they are at temperatures near to those at which they liquefy, it is found that they no longer conform to the ideal gas equation (nor, therefore, to the gas laws). This is not surprising because that equation is consistent with the kinetic theory of gases which assumes:

(i)   that there are no intermolecular forces, and

(ii) that the volume of the molecules is negligible compared with their separation.

An increase in pressure or a decrease in temperature clearly reduces the validity of assumption (ii). Assumption (i) also becomes less valid because there is ample evidence that the closely packed molecules of solids and liquids do exert forces on each other.

The extent of the departure from ideal gas behaviour varies from gas to gas, but of the common gases carbon dioxide shows considerable non-ideal characteristics.

7.7 ANDREWS' EXPERIMENTS ON CARBON DIOXIDE

The apparatus which Andrews used to investigate the behaviour of carbon dioxide is shown, schematically, in Fig. 7.6. By tightening the screws, Andrews was able to force water into the glass tubes and so increase the pressures and decrease the volumes of the gases trapped in the upper portions of the tubes. These tubes had been calibrated beforehand, so that it was a simple matter for Andrews to read off the volumes of the trapped gases by noting the positions of the tops of the mercury columns. By assuming that the nitrogen obeyed Boyle's law (a reasonable assumption at the pressures and temperatures involved as Andrews knew), he was able to calculate the pressure of the nitrogen once he had measured its volume. Since both gases were at the same pressure, this gave him the pressure of the carbon dioxide as desired. The capillary tubes were surrounded by a water bath, the purpose of which was to maintain the gases at a constant temperature. In this way then, Andrews measured the volume of the carbon dioxide as a function of its pressure at a fixed temperature. Altering the water bath temperature allowed him to obtain this information for a number of different temperatures. He presented his results as a series of **isothermals** (i.e. a series of plots of pressure against volume, each at a fixed temperature). Some of these curves are shown in Fig. 7.7.

Fig. 7.6
Andrews' apparatus

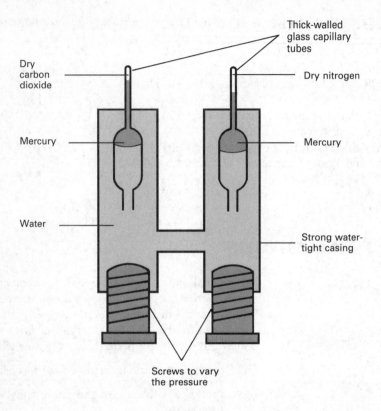

The diagram shows the critical nature of the 31.1 °C isothermal. Above 31.1 °C the carbon dioxide exists as a gas no matter how high the pressure, and the curves are approximately hyperbolic (the shape they would be if the carbon dioxide were an ideal gas). Below 31.1 °C the carbon dioxide can exist in both the gaseous state (as a vapour) and the liquid state. Consider the carbon dioxide to be in the state of pressure, volume and temperature that is represented by the point A on the 21.5 °C isothermal. In this state the carbon dioxide is an unsaturated vapour (see section 7.8), and if it is compressed, the *p–V* curve is very nearly hyperbolic until the pressure reaches that represented by B. At B the carbon dioxide begins to liquefy. Between B and C the volume decreases as the screws are turned in, but

Fig. 7.7
Andrews' isothermals
(*p–V* curves) for a fixed
mass of carbon dioxide

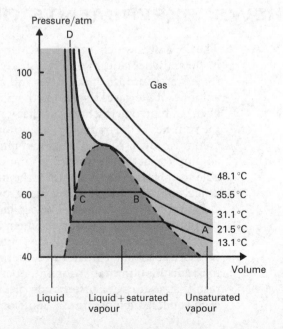

there is no increase in pressure. The decrease in volume is due to the fact that in moving from B to C more and more liquid forms, so that at C the carbon dioxide is entirely liquid. From C to D and beyond, large increases in pressure produce very little decrease in volume – as might be expected, since liquids are virtually incompressible.

7.8 TERMINOLOGY

It is now possible to define some useful terms.

Critical temperature (T_c) is the temperature above which a gas cannot be liquefied, no matter how great the pressure. $(T_c = 31.1\,°C$ for carbon dioxide.)

Critical pressure (p_c) is the minimum pressure that will cause liquefaction of a gas at its critical temperature. $(p_c = 73\,atm$ for carbon dioxide.)

Specific critical volume (V_c) is the volume occupied by 1 kg of a gas at its critical temperature and critical pressure.

Gas is the term applied to a substance which is in the gaseous phase and is above its critical temperature.

Vapour is the term applied to a substance which is in the gaseous phase and is below its critical temperature.

Thus, a vapour can be liquefied simply by increasing the pressure on it; a gas cannot. A vapour may be saturated or unsaturated. A **saturated vapour** is a vapour which is in equilibrium with its own liquid. Saturated vapours do not obey the gas laws. Unsaturated vapours obey the gas laws to the same extent as real gases and can be treated like an ideal gas when their temperatures and pressures are not close to those at which they would condense.

Notes (i) Oxygen, nitrogen and hydrogen are traditionally called **permanent gases**, since it was originally thought that they could not be liquefied. This misconception arose because the early workers had no knowledge of the necessity for a gas to be below its critical temperature, and each of these gases has a critical temperature which is well below room temperature $(-118\,°C, -146\,°C$ and $-240\,°C$ respectively).

(ii) It can be seen from the p–V curves of carbon dioxide (Fig. 7.7), for example, that when a liquid at its critical temperature (and critical pressure) becomes gaseous, then it does so without any change of volume. Under these conditions then, the liquid and its saturated vapour have the same density. Therefore, if a liquid and its saturated vapour are in equilibrium at their critical temperature, there is no meniscus, i.e. no distinction between liquid and vapour.

7.9 CURVES OF pV AGAINST p

A convenient way to show the departure from ideal gas behaviour at some temperature is to plot pV against p at that temperature. For an ideal gas such a plot is, of course, a straight line parallel to the p axis, but for a fixed mass of real gas the curves typically have the form shown in Fig. 7.8.

Fig. 7.8
Plots of **pV** against **p** for a
typical real gas

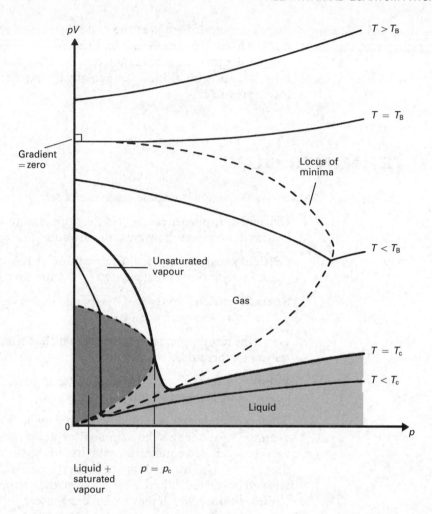

At the **Boyle temperature** T_B the isothermal is horizontal at $p = 0$, and therefore very nearly so for other low pressures, i.e. **the Boyle temperature of a gas is the temperature at which the gas approximates best to an ideal gas**.

Kamerlingh Onnes showed that the behaviour of all real gases can be accounted for by

$$pV = A + Bp + Cp^2 + \ldots^*$$ [7.4]

A, B, C, etc. are called **virial coefficients**, and for a fixed mass of any particular gas have values which depend <u>only on temperature</u>. On the whole, $|A| \gg |B| \gg |C|$, etc.

Two particular situations are of interest.

At $p = 0$

When the behaviour of any real gas is extrapolated to zero pressure it is found to behave like an ideal gas. Therefore, when $p = 0$, equation [7.4] must become identical with $pV = nRT$. Thus,

*Any function can be represented by an infinite polynomial such as equation [7.4], and therefore at first sight it might seem to be of little interest. The importance of the equation lies in:
(i)   each term is less significant than the preceding one, and
(ii) B varies in the same way with temperature for all gases.

$$A + Bp + Cp^2 + \cdots = nRT \quad \text{when } p = 0$$

i.e. $\boxed{A = nRT}$ [7.5]

Since the virial coefficients are not functions of p, equation [7.5] is true for all pressures

At the Boyle Temperature

At the Boyle temperature the graph of pV against p is horizontal at $p = 0$, i.e.

$$\frac{d(pV)}{dp} = 0 \qquad \text{when } p = 0$$

Therefore by equation [7.4]

$$\frac{d}{dp}(A + Bp + Cp^2 + \cdots) = 0 \quad \text{when } p = 0$$

i.e. $B + 2Cp + \cdots = 0 \qquad \text{when } p = 0$

i.e. $B = 0$

Thus at the Boyle temperature equation [7.4] becomes

$$pV = nRT + Cp^2 + \cdots$$

Since C and the succeeding coefficients are normally less significant than B (and B is actually zero here), to a good approximation, $pV = nRT$ at the Boyle temperature, as expected.

7.10 EVAPORATION

Evaporation is the process by which a liquid* becomes a vapour. It can take place at all temperatures, but occurs at the greatest rate when the liquid is at its boiling point.

The kinetic theory supposes that the molecules of liquids are in continual motion and make frequent collisions with each other. Although the average kinetic energy of a molecule is constant at any particular temperature, it may gain kinetic energy as a result of collisions with other molecules. If a molecule which is near the surface and is moving towards the surface gains enough energy to overcome the attractive forces of the molecules behind it, it escapes from the surface. It follows that the rate of evaporation can be increased by:

(i) increasing the area of the liquid surface;

(ii) increasing the temperature of the liquid (since this increases the average kinetic energy of all the molecules without increasing the strength of the intermolecular forces of attraction);

(iii) causing a draught to remove the vapour molecules before they have a chance to return to the liquid;

(iv) reducing the air pressure above the liquid (since this decreases the possibility of a vapour molecule rebounding off an air molecule).

*Solids evaporate but the rate of evaporation of a solid is negligible unless it is close to its melting point.

7.11 COOLING BY EVAPORATION

When a liquid evaporates it loses those of its molecules which have the greatest kinetic energies, and therefore **when a liquid evaporates it cools**.

This principle is made use of in the operation of a refrigerator. The **refrigerant** is a substance such as freon which can be liquefied at room temperature simply by being compressed, i.e. a substance whose critical temperature is above room temperature. The pump (Fig. 7.9) compresses the freon vapour and forces it through the condenser pipes. Cooling fins attached to the pipes remove the heat produced by the compression and the freon condenses. The liquid freon is forced through the valve into the evaporator unit on the low pressure side of the pump. The liquid evaporates, drawing its latent heat from the air and food in the refrigerator cabinet. The vapour is then compressed by the pump and the cycle repeats.

Fig. 7.9
The principle of the
refrigerator

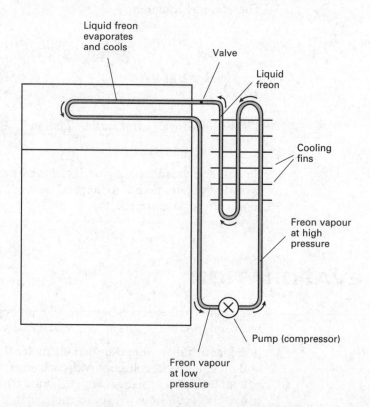

7.12 COOLING BY ADIABATIC EXPANSION OF A GAS

An adiabatic process is one which takes place in such a way that no heat enters or leaves the system during the process.

Suppose that a gas (real or ideal) expands adiabatically and does external work. In terms of the first law of thermodynamics (section 7.4), $\Delta Q = 0$ (because the expansion is adiabatic) and therefore $\Delta U = -\Delta W$ where ΔW is the work done by the gas and ΔU is the increase in its internal energy. Since the gas does do external work, $\Delta W > 0$ and therefore $\Delta U < 0$, i.e. the internal energy of the gas decreases. The decrease in internal energy causes a decrease in temperature (see Note (i)) and therefore **the temperature of a gas (real or ideal) decreases when it expands adiabatically and does external work**.

Notes (i)   The temperature of a gas is a measure of the <u>kinetic</u> energy of its molecules. The decrease in internal energy causes a decrease in molecular kinetic energy whether the gas is real or ideal. The internal energy of an ideal gas is entirely kinetic and therefore the decrease in kinetic energy is equal to the decrease in internal energy. The molecules of a real gas, on the other hand, have both potential energy and kinetic energy. The potential energy <u>increases</u> when the gas expands and therefore in the case of a real gas the decrease in kinetic energy is actually greater than the overall decrease in internal energy.

(ii)   It can be shown that when an ideal gas undergoes a reversible adiabatic expansion or contraction

$$p_1 V_1^{\gamma} = p_2 V_2^{\gamma} \qquad \text{[7.6]}$$

where p_1 and V_1 are the initial pressure and volume of the gas, p_2 and V_2 are the pressure and volume after the adiabatic change has taken place and γ is the ratio of the principal heat capacities of the gas*. If T_1 and T_2 are respectively the kelvin temperatures of the gas before and after the adiabatic change, then

$$\frac{p_1 V_1}{T_1} = \frac{p_2 V_2}{T_2} \qquad \text{[7.7]}$$

Dividing equation [7.6] by equation [7.7] gives

$$T_1 V_1^{(\gamma-1)} = T_2 V_2^{(\gamma-1)}$$

QUESTIONS 7B

1. An ideal gas at a temperature of 300 K undergoes a reversible adiabatic expansion from a volume of $1.20 \times 10^{-2}\,\text{m}^3$ to $6.00 \times 10^{-2}\,\text{m}^3$. Find the final temperature of the gas. ($\gamma = 1.40$).

2. The final pressure of the gas in Question 1 is $1.00 \times 10^5\,\text{Pa}$. What was its initial pressure?

7.13 THE JOULE–KELVIN EFFECT[†]

The Joule–Kelvin effect refers to the change in temperature that occurs when a <u>real</u> gas at some <u>constant</u> pressure is forced adiabatically through a narrow orifice into a region where it has some <u>constant</u> lower pressure.

Consider a gas at a constant pressure p_1 being forced adiabatically through a narrow orifice into a region where its pressure has the constant lower value p_2. Suppose also that a mass of the gas which has a volume V_1 at pressure p_1 occupies a volume V_2 when its pressure is p_2. The Joule–Kelvin effect may produce either an increase or a decrease in temperature. It is actually a combination of two separate effects. One of these always causes the temperature to decrease ((i) below); the other may cause either an increase or a decrease ((ii) below).

*See, for example, R. Muncaster, *A-Level Physics* (Stanley Thornes).
[†]The Joule–Kelvin effect is sometimes called the Joule–Thomson effect – Lord Kelvin was originally William Thomson.

(a) James Prescott Joule
    (1818–89)
(b) Lord Kelvin
    (1824–1907)

(i)   The gas expands on passing through the orifice (because $p_2 < p_1$) and therefore its molecules become further apart. This causes a decrease in temperature because work is done against the attractive forces between the molecules in order to increase their separation, and this has to be done at the expense of their kinetic energy since the process is carried out adiabatically.

(ii) It can be shown (see below) that the net work done on the gas in forcing it through the orifice is given by

$$\text{Net work done on gas} = p_1 V_1 - p_2 V_2$$

This may be positive or negative. If it is positive, work is done on the gas and its temperature increases; if it is negative, work is done by the gas and its temperature decreases.

Above the Boyle temperature

pV decreases with decreasing p (see Fig. 7.8) and therefore $p_1 V_1 - p_2 V_2 > 0$ and work is done on the gas. It follows that the overall effect may be either an increase or a decrease in temperature according to which of (i) and (ii) is the larger effect.

Below the Boyle temperature

For moderate pressures pV increases with decreasing p and therefore $p_1 V_1 - p_2 V_2 < 0$ and work is done by the gas. It follows that the overall effect in this case is bound to be a decrease in temperature because both (i) and (ii) produce a decrease.

> **The inversion temperature** of a gas is the temperature above which the Joule–Kelvin effect causes an increase in temperature and below which it causes a decrease.

From what has been said it should be obvious that **the inversion temperature of a gas is higher than its Boyle temperature**. At room temperature most gases are cooled by the Joule–Kelvin effect, but hydrogen and helium increase in temperature because their inversion temperatures are below room temperature (193 K and 30 K respectively).

Notes (i)   For an ideal gas $p_1 V_1 = p_2 V_2$ (Boyle's law) and there are no intermolecular forces. It follows that **there is no Joule–Kelvin effect with an ideal gas**.

(ii)   The change in temperature due to a Joule–Kelvin expansion is roughly proportional to the difference in pressure between the two sides of the orifice.

(iii)   One way of achieving a Joule–Kelvin expansion in practice is to pump a steady stream of gas through a nozzle or a porous plug into a region which is open to the atmosphere.

To show that work done on gas $= p_1 V_1 - p_2 V_2$

Fig. 7.10
To show that work done $= p_1 V_1 - p_2 V_2$

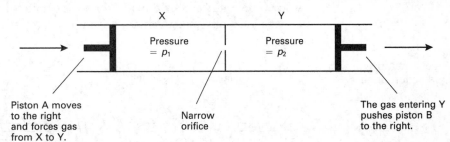

Piston A moves to the right and forces gas from X to Y.

Narrow orifice

The gas entering Y pushes piston B to the right.

Refer to Fig. 7.10. Suppose that piston A moves to the right and forces gas through the orifice from X to Y causing it to push piston B to the right in such a way that the pressure in X has the constant value p_1 and the pressure in Y has the constant value p_2. Suppose also that when the volume of gas in X decreases by V_1, that in Y increases by V_2. From equation [7.2]

Work done <u>on</u> gas by piston A $= p_1 V_1$

Work done <u>by</u> gas on piston B $= p_2 V_2$

∴   Net work done on gas $= p_1 V_1 - p_2 V_2$

7.14 LIQUEFACTION OF GASES BY JOULE–KELVIN EFFECT. THE COUNTERCURRENT HEAT EXCHANGER

Many gases (strictly, unsaturated vapours) can be liquefied simply by being compressed, but oxygen, nitrogen, hydrogen and helium have critical temperatures which are well below room temperature (see Table 7.1) and therefore have to be cooled considerably before they can be liquefied. The Joule–Kelvin effect can be used to cool a gas which is already below its inversion temperature. It therefore provides a means of cooling oxygen and nitrogen directly from room temperature (see Table 7.1). Unfortunately, hydrogen and helium have inversion temperatures

Table 7.1
Significant temperatures of some gases

Gas	Normal boiling point/K	Critical temperature/K	Inversion temperature/K
Oxygen	90	155	Above room temperature
Nitrogen	77	126	Above room temperature
Hydrogen	20	33	193
Helium	4.2	5.3	30

Note that the inversion temperature of any particular gas is much higher than its critical temperature.

which are below room temperature, and therefore have to be pre-cooled before the Joule–Kelvin effect can be employed. Hydrogen is cooled by passing it through a coil surrounded by (boiling) liquid nitrogen; helium is cooled by using liquid hydrogen.

A <u>single</u> Joule–Kelvin expansion does not produce sufficient cooling to cause a gas to liquefy and therefore a device called a **countercurrent heat exchanger** is used to make the effect cumulative in a process known as **regenerative cooling**.

Linde's method for the liquefaction of air is based on this principle. The apparatus is shown schematically in Fig. 7.11. The air which is to be liquefied is first passed through calcium chloride and caustic soda (not shown) to remove all traces of water vapour and carbon dioxide, for these would solidify and clog the system. The air is then compressed to about 200 atmospheres by a pump and passes through a

Fig. 7.11
Linde's method for the liquefaction of air

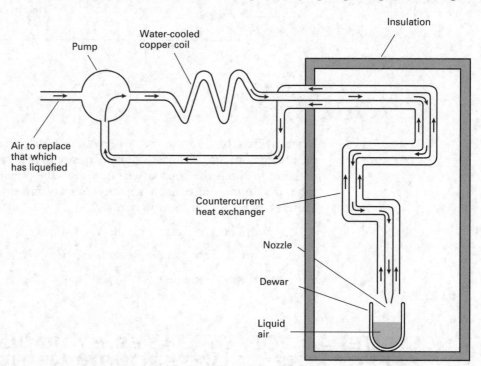

water-cooled copper coil to remove the heat generated by the compression. It then flows down the central section of a double-walled copper spiral (the countercurrent heat exchanger) and is forced out through a small nozzle, whereupon it expands and cools owing to the Joule–Kelvin effect. The cooled air, now at low pressure, passes back up the outer section of the double coil and therefore cools the incoming air, which on emerging from the nozzle becomes even colder. The cycle is repeated and eventually a steady state is reached in which a fraction of the air issuing from the nozzle liquefies and collects in the Dewer (vacuum) flask.

In order for the transfer of heat between the two opposing streams of air to be accomplished efficiently,

(i)   the inner wall of the countercurrent heat exchanger is thin and is made of copper (a good conductor),

(ii)   the air flow is at such a speed that it is turbulent and therefore all of the air comes into contact with the inner wall, and

(iii)   the double coil is relatively long.

Advantages of liquefying gases by using the Joule–Kelvin effect

(i) There are no moving parts which are at low temperatures – any such parts would be difficult to lubricate.

(ii) The lower the temperature, the greater the decrease in temperature for any given decrease in pressure.

CONSOLIDATION

The third law of thermodynamics   It will never be possible to reach absolute zero ($-273.15\,^{\circ}$C). (The third law is not unanimously accepted.)

Boyle's law   At constant T, $pV =$ a constant or $p \propto 1/V$

Charles' law   At constant p, $V/T =$ a constant or $V \propto T$

Pressure law   At constant V, $p/T =$ a constant or $p \propto T$

An ideal gas obeys $pV = nRT$ exactly. The internal energy (i.e. the energy of the molecules) is entirely kinetic and depends only on temperature.

A real gas obeys $pV = nRT$ approximately unless it is at high pressure and/or at a temperature close to that at which it liquefies. The molecules of a real gas exert forces on each other and therefore the internal energy of a real gas is both kinetic and potential.

$$W = \int_{V_1}^{V_2} pdV = \text{area under } p\text{–}V \text{ curve}$$

$$W = p(V_2 - V_1) \text{ at constant pressure}$$

First Law of Thermodynamics

$$\Delta Q = \Delta U + \Delta W$$

where

$$\Delta Q = \text{heat put \underline{into} system,}$$

$$\Delta U = \text{increase in internal energy of system,}$$

$$\Delta W = \text{work done \underline{by} system.}$$

$$\frac{p_1 V_1}{T_1} = \frac{p_2 V_2}{T_2} \quad \text{for any change of state of an ideal gas}$$

$$\left.\begin{aligned} p_1 V_1^{\gamma} &= p_2 V_2^{\gamma} \\ T_1 V_1^{(\gamma-1)} &= T_2 V_2^{(\gamma-1)} \end{aligned}\right\} \begin{aligned} &\text{for an \textbf{adiabatic} (i.e. constant heat)} \\ &\text{change of an ideal gas} \end{aligned}$$

The critical temperature of a gas is the temperature above which the gas cannot be liquefied, no matter how great the pressure.

The Boyle temperature of a gas is the temperature at which it approximates best to an ideal gas.

When a liquid evaporates it cools.

The temperature of both real and ideal gases decreases in an adiabatic expansion in which work is done by the gas.

The Joule–Kelvin effect refers to the change in temperature that occurs when a real gas at some <u>constant</u> pressure is forced adiabatically through a narrow orifice into a region where it has some <u>constant</u> lower pressure. There is an increase in temperature if the gas is above its **inversion temperature** and a decrease if it is below it.

There is no Joule–Kelvin effect for an ideal gas.

The Joule–Kelvin effect can be used to liquefy gases. Hydrogen and helium need to be pre-cooled because their inversion temperatures are above room temperature.

QUESTIONS ON CHAPTER 7

1. At a temperature of 100 °C and a pressure of 1.01×10^5 Pa, 1.00 kg of steam occupies 1.67 m^3 but the same mass of water occupies only 1.04×10^{-3} m^3. The specific latent heat of vaporization of water at 100 °C is 2.26×10^6 J kg^{-1}. For a system consisting of 1.00 kg of water changing to steam at 100 °C and 1.01×10^5 Pa, find:
   (a) the heat supplied to the system,
   (b) the work done by the system,
   (c) the increase in internal energy of the system.   [C]

2. (a) The first law of thermodynamics is represented by the equation

$$Q = \Delta U + W$$

   Explain each term in this equation.
   (b) An engine (shown below) burns a mixture of petrol vapour and air. When the engine is running it makes 25 power strokes per second and develops a mean power of 18 kW.

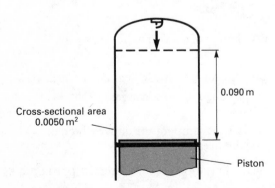

Cross-sectional area
0.0050 m^2

0.090 m

Piston

   Neglecting losses in the engine due to friction and other causes, calculate the work done in each power stroke.

   (c) The burning starts when the piston is at the top of its stroke and the resulting high pressure drives the piston downwards through a distance of 0.090 m. The cylinder has a cross-sectional area of 0.0050 m^2. Calculate:
      (i) the mean force on the piston head during the power stroke;
      (ii) the mean pressure of the hot gas.
                                             [O, '92]

3. Why is the energy needed to raise the temperature of a given mass of gas by a certain amount greater if the pressure is kept constant than if the volume is kept constant?   [L]

4. The specific latent heat of vaporization of a particular liquid at 130 °C and a pressure of 2.60×10^5 Pa is 1.84×10^6 J kg^{-1}. The specific volume of the liquid under these conditions is 2.00×10^{-3} m^3 kg^{-1}, and that of the vapour is 5.66×10^{-1} m^3 kg^{-1}. Calculate:
   (a) the work done, and
   (b) the increase in internal energy when 1.00 kg of the vapour is formed from the liquid under these conditions.

5. (a) Explain what is meant by a *reversible* change.
   (b) State the *first law of thermodynamics*, and discuss the experimental observations on which it is based.
   (c) A mass of 0.35 kg of ethanol is vaporized at its boiling point of 78 °C and a pressure of 1.0×10^5 Pa. At this temperature, the specific latent heat of vaporization of ethanol is 0.95×10^6 J kg^{-1}, and the densities of the liquid and vapour are 790 kg m^{-3} and 1.6 kg m^{-3} respectively.

Calculate:
(i) the work done by the system;
(ii) the change in internal energy of the system.
Explain in molecular terms what happens to the heat supplied to the system.   [O]

6. Describe briefly the experiments which Andrews performed on carbon dioxide. (A detailed description of the apparatus is *not* required.)
   (a) Draw graphs to show the pressure–volume relationship which Andrews obtained for various temperatures. Indicate on your diagram the various states of the carbon dioxide.
   (b) Use your graphs to explain the meaning of critical temperature. What is its significance in connection with the liquefaction of gases?   [AEB, '79]

7. **(a)** State the conditions under which the behaviour of a *real* gas will deviate significantly from that expected of an *ideal* gas.
   (b) (i) On a pV against p diagram sketch an isotherm for a real gas at the Boyle temperature. On the same set of axes sketch isotherms for temperatures just above and just below the Boyle temperature, labelling the isotherms clearly.
   (ii) Explain how the properties of the atoms or molecules of a gas give rise to the shape of the isotherm you have drawn *below* the Boyle temperature.
   (c) A quantity of oxygen gas occupies $0.20\,\mathrm{m}^3$ at a temperature of $27\,^\circ\mathrm{C}$ and pressure of 10 atmospheres. If it were to be liquefied, what volume of liquid oxygen, density $1.1 \times 10^3\,\mathrm{kg\,m}^{-3}$, would be produced? The oxygen gas in its initial state may be considered to behave as an ideal gas.
   What condition must be met before the gas can be liquefied by the increase of pressure alone?
   (1 atmosphere $= 1.0 \times 10^5\,\mathrm{Pa}$, relative molecular mass of oxygen $= 32$, molar gas constant $= 8.3\,\mathrm{J\,mol}^{-1}\,\mathrm{K}^{-1}$.)   [J]

8. **(a)** The partly labelled diagram below shows the apparatus used by Andrews in his experiments on carbon dioxide.
   State what A is and explain its purpose.
   Explain why each of the water baths, P, Q and R were used during the experiment.

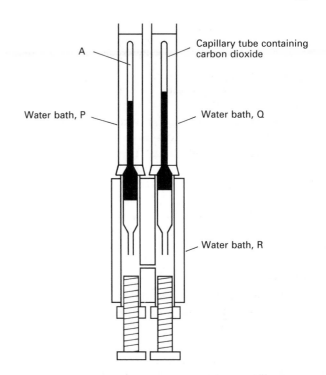

(b) State the meanings of *critical temperature* and *critical pressure*.
(c) Some carbon dioxide initially at a temperature above its critical temperature is subjected to the following changes.
   (i) It is compressed isothermally to a pressure above its critical pressure.
   (ii) Then at this pressure it is cooled at constant pressure until the temperature is well below its critical temperature.
   (iii) Then at this temperature it is expanded isothermally until all the carbon dioxide is again a gas.
Sketch a graph of pressure against volume to illustrate these changes, and discuss the associated changes of state.   [J]

9. **(a)** Sketch isothermal curves to show how the pressure of a fixed mass of substance (e.g. carbon dioxide) varies with volume over a wide range of temperature and pressure. Indicate on your sketch the regions where the substance is in *the liquid phase, the saturated vapour phase, the unsaturated vapour phase* and *the gas phase*.
   (b) An unsaturated vapour of mass $5 \times 10^{-4}\,\mathrm{kg}$ and at a temperature of $20\,^\circ\mathrm{C}$ is compressed isothermally until, at a volume $V_1 = 9 \times 10^{-5}\,\mathrm{m}^3$ and a pressure $6 \times 10^6\,\mathrm{Pa}$, the vapour first becomes saturated. Further compression of the vapour

causes the formation of liquid until, when the volume is V_2, the substance is changed completely to liquid. If V_2 is negligible compared with V_1 and the temperature remains constant throughout the process, calculate:

(i) the work that must be performed during the compression from V_1 to V_2,

(ii) the amount of thermal energy that must be supplied to, or removed from, the substance during the same compression.

(Assume that the specific latent heat of vaporization of the liquid at $20\,°C$ is $1.2 \times 10^5\,J\,kg^{-1}$.)   [AEB, '79]

10. Explain:

(a) why a liquid cools when it evaporates, and

(b) how this is made use of in the operation of a refrigerator.

11. What is an *adiabatic* change?

A vessel of volume $8.00 \times 10^{-3}\,m^3$ contains an ideal gas at a pressure of $1.14 \times 10^5\,Pa$. A stopcock in the vessel is opened and the gas expands adiabatically, expelling some of its original mass, until its pressure is equal to that outside the vessel ($1.01 \times 10^5\,Pa$). The stopcock is then closed and the vessel is allowed to stand until the temperature returns to its original value; in this equilibrium state, the pressure is $1.06 \times 10^5\,Pa$.

(a) Explain why there was a temperature change as a result of the adiabatic expansion.

(b) Find the volume which the mass of gas finally left in the vessel occupied under the original conditions.

(c) Sketch a graph showing the way in which the pressure and volume of the mass of gas finally left in the vessel changed during the operations described above. (It is *not* necessary to plot exact numerical values of p and V.)

(d) What is the value of γ, the ratio of the principal heat capacities of the gas?   [C]

12. Explain what is meant by the Joule–Kelvin effect.

13. What are the essential differences between the cooling of a gas by adiabatic expansion and by the Joule–Kelvin effect?

14. Explain what is meant by the terms, *critical temperature, Boyle temperature* and *inversion temperature*. Which is the highest temperature of these three for any particular gas?

15. Why were Andrews' experiments on carbon dioxide of crucial importance to attempts to liquefy the so-called permanent gases?

16. (a) Describe, with the aid of a diagram, Linde's method for the liquefaction of air.

(b) Explain the modification that must be made if the method is to be used to liquefy hydrogen.

17. (a) Outline the principle of a countercurrent heat exchanger used to liquefy air.

(b) Why is it necessary to pre-cool hydrogen gas before it can be liquefied using a countercurrent heat exchanger?

[J (specimen), '96]

8

SUPERCONDUCTIVITY AND SUPERFLUIDITY

8.1 SUPERCONDUCTIVITY

Superconductivity was discovered by Heike Kamerlingh Onnes in 1911 whilst investigating the electrical resistivity of mercury at temperatures close to absolute zero. It is characterized by the <u>sudden</u> disappearance of electrical resistance when the temperature is reduced below some **critical temperature***, T_c, (Fig. 8.1). The critical temperature depends on the material concerned; for mercury it is 4.15 K.

Fig. 8.1
Variation of resistance with temperature for a typical superconductor

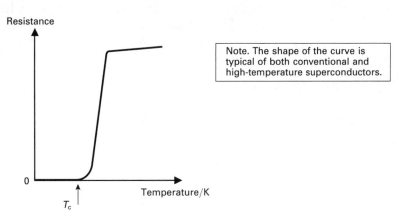

Note. The shape of the curve is typical of both conventional and high-temperature superconductors.

Materials which exhibit superconductivity are called **superconductors**. The reader should be clear that the resistance of a superconductor below its critical temperature really is <u>zero</u>. Because the resistance is zero, once a current has been established in a superconductor, it will flow for ever without any applied PD and without producing any heat. Such currents are known as **persistent currents** or **supercurrents**. In an experiment started in 1955, a current <u>induced</u> in a superconducting lead ring was observed to flow for over two years without any detectable decrease in strength. The current stopped only when the temperature was allowed to rise above the critical temperature of lead.

Many of the <u>metallic</u> elements are superconductors. Notable amongst those that are not are copper, silver and gold (all of which are excellent conductors in the normal sense) and the ferromagnetic elements iron, cobalt and nickel. Of the elemental superconductors, niobium has the highest critical temperature, 9.3 K. Thousands of alloys and compounds are also known to exhibit superconductivity.

*This should not be confused with the critical temperature of a gas referred to in Chapter 7.

Notes (i) The phenomenon of superconductivity owes its name to its most obvious (and first to be discovered) characteristic – the complete absence of electrical resistance at temperatures below the critical temperature. However, a number of other properties also change markedly at the critical temperature. For example, if a superconductor is cooled to its critical temperature:

(a) all thermoelectric effects disappear,

(b) there is a sudden change in specific heat capacity, and

(c) the superconductor becomes perfectly diamagnetic and because of this it can levitate a permanent magnet placed above it.

(ii) Superconductors are classified as being either **type I** superconductors or **type II**, according to how they behave under the influence of an applied magnetic field. The bulk of the elemental metals which exhibit superconductivity are type I (niobium is an exception). Most of the superconducting alloys and compounds, on the other hand, are type II.

8.2 CRITICAL MAGNETIC FIELD AND CRITICAL CURRENT DENSITY

If a magnetic field is applied to a superconductor, it ceases to be superconducting if the flux density exceeds the so-called **critical field**, B_c. The value of B_c increases with decreasing temperature, ranging from zero at $T = T_c$ to its maximum value, $B_c(0)$, when $T = 0\,\text{K}$ (Fig. 8.2). The critical temperatures, T_c, and maximum

Fig. 8.2
Variation of critical field strength with temperature for a superconductor

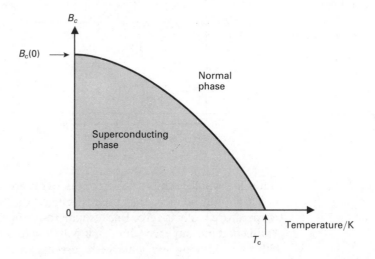

critical fields, $B_c(0)$, of some selected superconducting materials are listed in Table 8.1. Note that the alloys have much higher values of $B_c(0)$ than the elemental metals and so can be used for manufacturing superconducting electromagnets.

Superconductivity is also destroyed if the current density exceeds a critical value known as the **critical current density**, J_c. The value of J_c varies in much the same way as that of B_c, increasing from zero at $T = T_c$ to its maximum value, $J_c(0)$, at $T = 0\,\text{K}$. Materials which are used for superconducting electromagnets require high values of both $B_c(0)$ and $J_c(0)$. One such material is Nb_3Sn which has $B_c(0) = 22\,\text{T}$ and $J_c(0) \sim 10^{10}\,\text{A m}^{-2}$.

Table 8.1
Critical temperatures and
critical fields of some
superconductors

Superconductor		T_c/K	$B_c(0)/T*$
Zn		0.85	5.4×10^{-3}
Sn	Type I	3.72	3.1×10^{-2}
Hg		4.15	4.1×10^{-2}
Pb		7.19	8.0×10^{-2}
Nb		9.26	4.0×10^{-1}
PbMoS	Type II	14.4	59
Nb_3Sn		18.1	22
Nb_3Ge		23.3	38

*Type II superconductors have two critical fields –
upper and lower. Though there is a marked change
in the magnetic properties of a superconductor
when its lower critical field is exceeded, the
electrical resistivity remains zero until the upper
field is exceeded. The values of $B_c(0)$ listed here
are those of the upper critical field.

Note A superconductor which is carrying a current has an associated magnetic field. If
the flux density of this field at the surface of the conductor exceeds the critical field,
B_c, the material ceases to be superconducting. Thus the maximum current that a
superconductor can carry is limited by the diameter of the conductor and the
critical field of the material concerned. This is quite distinct from the limitation
imposed by the critical current density.

EXAMPLE 8.1

The critical field of lead at the boiling point of liquid helium (4.2 K) is
5.3×10^{-2} T. What is the maximum current that can flow in a lead wire with a
radius of 4.0 mm if it is to remain superconducting at this temperature?

Solution

The flux density, B, at a distance r from the centre of a wire carrying a current I is
given by

$$B = \frac{\mu_0 I}{2\pi r}$$

where μ_0 is the permeability of vacuum $(= 4\pi \times 10^{-7}\,H\,m^{-1})$. The flux density at
the surface of the wire is therefore given by

$$B = \frac{4\pi \times 10^{-7} I}{2\pi \times 4.0 \times 10^{-3}} = 5.0 \times 10^{-5} I$$

Since this must not exceed 5.3×10^{-2} T

$$5.0 \times 10^{-5} I \leqslant 5.3 \times 10^{-2}$$

i.e. $I \leqslant 1.06 \times 10^3$

i.e.   Maximum current $\approx 1.1 \times 10^3$ A

8.3 HIGH-TEMPERATURE SUPERCONDUCTORS

For many years the highest known critical temperature was 23.3 K, (for a niobium–germanium alloy, Nb_3Ge). However, in 1986 a ceramic material (a Ba–La–Cu–O compound) was discovered that had a critical temperature of 35 K. This created much interest in ceramics as superconductors and early in 1987 the discovery of a material with a critical temperature of 92 K was announced. This was a major discovery because it meant that liquid nitrogen (boiling point 77 K) could be used to achieve the cooling required for the onset of superconductivity. Prior to this it had been necessary to use liquid helium (which is expensive) or liquid hydrogen (which is explosive). Liquid nitrogen is cheap and is much easier to use than either helium or hydrogen.

Research continues in the hope of eventually finding materials which are superconducting at room temperature. To date, the highest known critical temperature is 125 K. All the high-temperature superconductors so far discovered are copper oxide compounds.

8.4 APPLICATIONS OF SUPERCONDUCTIVITY

The discovery of the high-temperature superconductors opens up the possibility of superconducting devices operating at liquid nitrogen temperatures or, as more are discovered, perhaps even at room temperature. However, because they are ceramics, they are brittle and are not easily produced in the form of wire. Another problem is that although the critical current density measured along certain lattice directions in thin films of these materials is high, that of bulk specimens is much lower than that of alloys such as Nb_3Sn. If these problems can be overcome, there will be many exciting possibilities, some of which are discussed below.

Superconducting Electromagnets

Superconducting electromagnets are now routinely used in high-energy particle accelerators. They are capable of much higher flux densities (≈ 10 T) than conventional electromagnets, which are limited by the need to remove the heat generated in the coils as a result of resistive losses. Furthermore, because no heat is generated in the coils of a superconducting electromagnet, there is no need to incorporate channels through which cooling fluids can flow and therefore they can be made much more compact than conventional magnets. Superconducting magnets consume no power other than that required to run the cooling system. Those currently in use operate at liquid helium temperatures. The high-temperature superconductors offer the possibility of using liquid nitrogen as the coolant which would reduce operating costs still further.

The so-called body scanners used as a diagnostic aid in medicine also make use of superconducting magnets. The development of magnets operating at liquid nitrogen temperatures would also produce significant cost savings in this area.

Power Transmission

The heat generated through resistive losses in conventional high-tension transmission lines wastes about 10% of the power supplied. This would be eliminated by the use of superconducting cables.

Magnetic Levitation

If a superconductor is placed on a permanent magnet and then cooled below its critical temperature, it rises off the magnet and remains suspended above it. This magnetic levitation effect has already been exploited by the Japanese in the construction of a prototype train which has achieved speeds in excess of 300 mph suspended a few inches above its tracks. The prototype uses liquid helium as the coolant but vehicles of this type might become commonplace if liquid nitrogen could be used instead.

(a) Permanent magnet levitated above a pellet of $YBa_2Cu_3O_{7-x}$ superconductor cooled to liquid nitrogen temperature.
(b) Prototype magnetically levitated train built in Japan

8.5 SUPERFLUIDITY

At temperatures below 2.17 K liquid helium 4 behaves as if it is a mixture of two fluids – a normal fluid and a **superfluid**. The viscosity of the superfluid component is <u>zero</u> and it can flow through capillaries which are so narrow that even gases cannot pass through them.

The ability of liquid helium to flow with such remarkable ease was discovered by P. L. Kapitza in 1938, and is just one of a number of unusual properties exhibited by liquid helium 4 when it is below 2.17 K and which, collectively, are known as **superfluidity**. Some examples are given below.

Almost Perfect Thermal Conductivity

Liquid helium 4 has exceptionally high thermal conductivity at temperatures below 2.17 K. This was discovered by Keesom and Keesom in 1935 and is what prompted Kapitza to make his more detailed investigation of the liquid three years later. The Keesoms (father and daughter) cooled liquid helium 4 below its normal boiling point of 4.2 K by reducing the pressure on it. They noticed that the bubbles which are characteristic of a boiling liquid suddenly stopped forming once the temperature had fallen below 2.17 K. This indicated that the liquid was now an almost perfect conductor of heat – there were no bubbles because all parts of the liquid were always at the same temperature. It is, in fact, the best conductor of heat known; its thermal conductivity is about 10^6 times that of its value above 2.17 K. Perhaps surprisingly though, it remains an electrical insulator throughout.

Specific Heat Capacity

There is a marked change in the specific heat capacity of liquid helium 4 at 2.17 K (see Fig. 8.3).

Fig. 8.3
Specific heat capacity of
liquid helium 4 as a
function of temperature

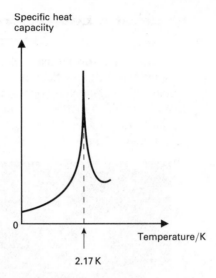

Thermomechanical Effect

If heat is applied to liquid helium 4 which is below 2.17 K, the superfluid
component of the liquid flows towards the source of the heat – **the
thermomechanical effect**. In Fig. 8.4(a) the liquid levels in the two containers
are initially the same. When the heater is switched on (Fig. 8.4(b)) liquid passes
through the narrow tube towards the heat source and raises the level in the inner
container – the narrower the tube, the greater the effect. This forms the basis of a
striking demonstration known as the **fountain effect** (Fig. 8.5). The tube below
the heater is packed with a fine powder creating lots of very narrow capillaries.
When the heater is switched on liquid helium squirts out of the top of the tube in a
fountain which has been observed as much as 100 cm in height for a temperature
difference of less than 0.1 °C.

Fig. 8.4
The thermomechanical
effect

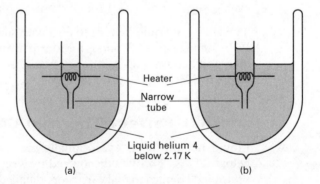

The converse also occurs. Refer to Fig. 8.6. Increasing the pressure in side A forces
liquid from A to B. The temperature of the liquid in A increases slightly and that in
B decreases. This is because it is the superfluid component that passes through the
capillary tube, and the atoms of the superfluid are in a lower energy state than those
of the normal fluid.

Film Creep

Liquid helium 4 below 2.17 K can escape from an open-topped container by
creeping, in the form of a thin film, up the walls and over the rim. (The film is
estimated to be fewer than 100 atoms thick.) In each of Fig. 8.7(a) and (b) the
liquid creeps in the directions of the arrows until the levels in the two containers are
equal. In Fig. 8.7(c) the process continues until the inner container is empty.

Fig. 8.5
The fountain effect

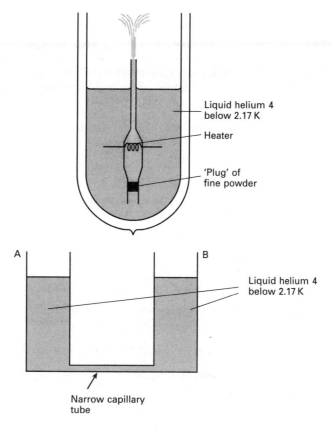

Liquid helium 4
below 2.17 K

Heater

'Plug' of
fine powder

Fig. 8.6
Apparatus to
demonstrate the thermo-
mechanical effect with
liquid helium 4

Liquid helium 4
below 2.17 K

Narrow capillary
tube

Notes

(i) The temperature at which the transition to the superfluid phase occurs is called the **lambda point**, T_λ, because (around this value) the graph of specific heat capacity versus temperature resembles the Greek letter λ.

(ii) Liquid helium 4 is known as **helium I** above T_λ and as **helium II** below it.

(iii) The ratio of superfluid to normal fluid increases as temperature decreases; the liquid is almost all superfluid at temperatures below 1 K.

(iv) Superfluidity (in helium 4) is due to a quantum-mechanical phenomenon known as **Bose–Einstein condensation** – it cannot be explained on the basis of classical physics.

(v) Though the electrons in a superconductor can be regarded as a superfluid, the only <u>liquid</u> other than helium 4 which is known to exhibit superfluidity is liquid helium 3, which becomes superfluid at about 0.003 K.

Fig. 8.7
Demonstration of the
creeping film effect

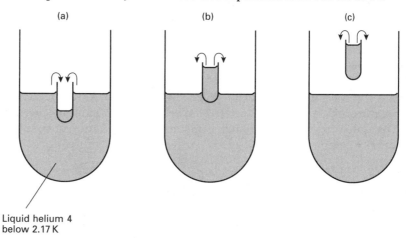

(a) (b) (c)

Liquid helium 4
below 2.17 K

CONSOLIDATION

A superconductor is a material whose electrical resistivity is zero when it is below its **critical temperature**. There are also marked changes in other properties (e.g. specific heat capacity, and magnetic and thermoelectric behaviour) at the critical temperature.

Once a current has been established in a superconductor, it will flow for ever unless the temperature rises above T_c. Such currents are called **persistent currents** or **supercurrents**.

Superconductors cease to be superconducting if the temperature exceeds T_c or the current density exceeds J_c or if they are in a magnetic field (either external or due to a current in the superconductor itself) of flux density greater than B_c.

Superconducting electromagnets are superior to conventional electromagnets in that they are capable of producing higher flux densities, are more compact and are cheaper to run. If magnets operating at liquid nitrogen temperatures are developed, they will be even cheaper.

Superfluidity – the ability of liquid helium to flow with zero viscosity and to exhibit other unusual effects.

Helium I – liquid helium 4 between 4.2 K and 2.17 K.

Helium II – liquid helium 4 below 2.17 K.

The lambda point (T_λ) is the temperature below which liquid helium 4 exhibits superfluidity.

QUESTIONS ON CHAPTER 8

1. What is meant by **(a)** the critical temperature, **(b)** the critical magnetic field of a super-conductor?

2. What are persistent currents?

3. Certain compounds have been found to be superconducting at temperatures greater than 77 K, the boiling point of liquid nitrogen. Sketch a graph to show how you would expect the resistivity of such a compound to vary from 77 K to above its superconducting critical temperature.   [J (specimen), '96]

4. Explain the ways in which the so-called high-temperature superconductors are **(a)** superior to, **(b)** inferior to conventional superconductors.

5. What is the maximum current that can flow in a Nb_3Al wire of radius 2.4 mm at 4.2 K if the wire is to remain superconducting? (The (upper) critical field of Nb_3Al at 4.2 K is 31 T.)

6. A magnetic field of flux density 1.8×10^{-2} T is perpendicular to the plane of a supercon-ducting ring of radius 1.2 cm. When the field is reduced to zero a current of 200 A is induced in the ring. Calculate the inductance of the ring. (Note. Change in flux = Inductance × Induced current.)

7. At temperatures below 2.17 K liquid helium 4 exhibits a number of effects collectively known as superfluidity. Give details of three of these.

VALUES OF SELECTED PHYSICAL CONSTANTS

Quantity	Symbol	Value
Speed of light in vacuum	c	$2.99792 \times 10^8 \, \text{m s}^{-1}$
Planck's constant	h	$6.62608 \times 10^{-34} \, \text{J s}$
Electronic charge	e	$1.60218 \times 10^{-19} \, \text{C}$
Mass of electron	m_e	$9.10939 \times 10^{-31} \, \text{kg}$
		$5.48580 \times 10^{-4} \, \text{u}$
		$5.10999 \times 10^{-1} \, \text{MeV}/c^2$
Mass of proton	m_p	$1.67262 \times 10^{-27} \, \text{kg}$
		$1.00728 \, \text{u}$
		$9.38272 \times 10^2 \, \text{MeV}/c^2$
Mass of neutron	m_n	$1.67493 \times 10^{-27} \, \text{kg}$
		$1.00867 \, \text{u}$
		$9.39566 \times 10^2 \, \text{MeV}/c^2$
Unified atomic mass unit	u	$1.66054 \times 10^{-27} \, \text{kg}$
		$9.31494 \times 10^2 \, \text{MeV}/c^2$
Avogadro constant	N_A	$6.02214 \times 10^{23} \, \text{mol}^{-1}$
Permittivity of vacuum	ε_0	$8.85419 \times 10^{-12} \, \text{F m}^{-1}$
	$1/4\pi\varepsilon_0$	$8.98755 \times 10^9 \, \text{m F}^{-1}$
Boltzmann's constant	k	$1.38066 \times 10^{-23} \, \text{J K}^{-1}$

ANSWERS TO END OF CHAPTER QUESTIONS

The Examination Boards accept no responsibility whatsoever for the accuracy or method of working in the answers given. These are the responsibility of the author.

CHAPTER 1

1. $9.4 \times 10^7 \, \text{m s}^{-1}$
2. (a) (i) $3.8 \times 10^7 \, \text{m s}^{-1}$ (ii) $6.0 \times 10^{-15} \, \text{N}$
   (b) 21 cm
3. (a) (ii) $2.8 \times 10^2 \, \text{V}$
4. (c) (i) $8.3 \times 10^{-7} \, \text{s}$ (ii) $4.3 \times 10^{-5} \, \text{T}$
5. $1.33 \times 10^7 \, \text{m s}^{-1}$, $1.71°$
6. $6.7 \times 10^7 \, \text{m s}^{-1}$
8. (a) $4.2 \times 10^7 \, \text{m s}^{-1}$ (b) 0.12 m
10. (b) (i) $6.0 \times 10^4 \, \text{V m}^{-1}$ (ii) $1.6 \times 10^{-19} \, \text{C}$
11. (b) $41 \, \text{kV m}^{-1}$
13. 6
14. (a) (iv) $3.6 \times 10^{-15} \, \text{N}$
15. $6.4 \times 10^{-19} \, \text{C}$
16. $4.68 \times 10^{-19} \, \text{C}$
17. $4.8 \times 10^{-19} \, \text{C}$; $1.6 \times 10^{-19} \, \text{C}$

CHAPTER 2

6. (b) $3.1 \times 10^4 \, \text{m}$, $300 \, \text{rev s}^{-1}$
8. $3.0(2) \times 10^8 \, \text{m s}^{-1}$

CHAPTER 3

3. 1073 K
6. (a) 0.19 nm   (b) 1.07 mm
9. 378 K
10. (a) $5.76 \times 10^3 \, \text{K}$ (b) $5.03 \times 10^{-7} \, \text{m}$
14. $1.96 \times 10^{-19} \, \text{J}$
15. $1.96 \times 10^{-19} \, \text{J}$, $> 1.01 \times 10^{-6} \, \text{m}$
16. 4.45 eV
17. 1.5 V, $2.5 \times 10^{-19} \, \text{J}$, $7.3 \times 10^5 \, \text{m s}^{-1}$
18. $6.64 \times 10^{-34} \, \text{J s}$
19. (c) (i) $2.35 \times 10^5 \, \text{m s}^{-1}$ (ii) 0.157 V
20. (b) (i) $3.6 \times 10^{-19} \, \text{J}$ (ii) $2.8 \times 10^{17} \, \text{s}^{-1}$ (iii) 9.1%
   (c) (i) $2.2 \times 10^{-11} \, \text{m}$
21. (b) $6.7 \times 10^{-34} \, \text{J s}$ (c) (i) no emission
   (ii) electrons emitted with a KE of $3.74 \times 10^{-19} \, \text{J}$
22. (a) 1 eV   (b) (i) 1 V   (ii) 0.75 V
23. (c) (i) $3.4 \times 10^{-19} \, \text{J}$ (ii) $6.68 \times 10^{-34} \, \text{J s}$
24. (b) (ii) (I) $2.6 \times 10^{-19} \, \text{J}$ (II) $7.5 \times 10^{-7} \, \text{m}$
   (c) $1.24 \times 10^{-6} \, \text{m}$
25. $2.5 \times 10^{17} \, \text{s}^{-1}$, 3.9 mA, 0.39 V
26. (a) (i) $1.6 \times 10^{-9} \, \text{A}$ (ii) $3.2 \times 10^{-9} \, \text{A}$ (iii) zero
   (b) 0.5 V
27. (b) (i) 1.0 V   (ii) $1.6 \times 10^{-19} \, \text{J}$ (iii) $3.8 \times 10^{-19} \, \text{J}$

CHAPTER 4

1. $1.9 \times 10^{-10} \, \text{m}$
2. (a) $8.5 \times 10^{-36} \, \text{m}$ (b) $3.3 \times 10^{-38} \, \text{m}$
4. $7.3 \times 10^6 \, \text{m s}^{-1}$
5. (a) 43:1   (b) $1.9 \times 10^3 : 1$
6. (a) (ii) $2.7 \times 10^7 \, \text{m s}^{-1}$
7. (a) $7.09 \times 10^{-12} \, \text{m}$

CHAPTER 5

1. (a) 661 nm, 489 nm, 436 nm, 412 nm
2. (a) $3.4 \times 10^{-19} \, \text{J}$ (b) $1.8 \times 10^{20} \, \text{s}^{-1}$
3. $1.22 \times 10^{-7} \, \text{m}$
4. (a) $2.16 \times 10^{-18} \, \text{J}$ (b) $6.6 \times 10^{-7} \, \text{m}$
5. (b) (ii) $5.1 \times 10^{-19} \, \text{J}$ (iii) $3.9 \times 10^{-7} \, \text{m}$
6. (a) (i) 1.9 eV   (ii) 10.2 eV
7. (a) (i) $3.3 \times 10^{15} \, \text{Hz}$ (ii) $2.5 \times 10^{15} \, \text{Hz}$
9. (a) $2.9 \times 10^{15} \, \text{Hz}$, $2.5 \times 10^{15} \, \text{Hz}$, $4.6 \times 10^{14} \, \text{Hz}$
   (b) none
10. (b) (vi) $1.1 \times 10^7 \, \text{m}^{-1}$
11. $3.8 \times 10^{15} \, \text{s}^{-1}$

CHAPTER 6

3. 43.2 ns
4. (a) $0.8c$ (b) 10 years
5. $0.95c$
7. $5 \times 10^{-3} \, \%$
8. $1.5 \times 10^4 \, \text{kg m}^{-3}$
9. (a) 0.80   (b) 0.72
10. (a) $1.2 \times 10^{-12} \, \text{m}$
13. (a) $1.64 \times 10^{-13} \, \text{J}$ (b) $1.24 \times 10^{20} \, \text{Hz}$
15. 0.75%
16. $\frac{1}{3}c\sqrt{5}$ ($\approx 0.745c$)
18. $0.98c$
19. (a) $8.15 \times 10^7 \, \text{m s}^{-1}$ (b) $7.71 \times 10^{-23} \, \text{kg m s}^{-1}$
20. 7.7 cm

CHAPTER 7

1. (a) $2.26 \times 10^6 \, \text{J}$ (b) $1.69 \times 10^5 \, \text{J}$ (c) $2.09 \times 10^6 \, \text{J}$
2. (b) $7.2 \times 10^2 \, \text{J}$ (c) (i) 8.0 kN   (ii) 1.6 MPa
4. (a) $1.47 \times 10^5 \, \text{J}$ (b) $1.69 \times 10^6 \, \text{J}$
5. (c) (i) $2.2 \times 10^4 \, \text{J}$ (ii) $3.1 \times 10^5 \, \text{J}$
7. (c) $2.3 \times 10^{-3} \, \text{m}^3$
9. (b) (i) 540 J   (ii) 60 J
11. (b) $7.44 \times 10^{-3} \, \text{m}^3$ (d) 1.66

CHAPTER 8

5. $3.7 \times 10^5 \, \text{A}$
6. $4.1 \times 10^{-8} \, \text{H}$

ANSWERS TO QUESTIONS 1A–7B

QUESTIONS 1A

1. $8.4 \times 10^5 \, \mathrm{m\,s^{-1}}$
2. (a) $1.3 \times 10^6 \, \mathrm{m\,s^{-1}}$ (b) $5.4 \times 10^5 \, \mathrm{m\,s^{-1}}$

QUESTIONS 1B

1. $8.5 \, \mathrm{cm}$
2. (a) $5.0 \times 10^{-8} \, \mathrm{s}$ (b) $5.3 \times 10^{13} \, \mathrm{m\,s^{-2}}$
   (c) $6.6 \, \mathrm{cm}$ (d) $3.3 \times 10^6 \, \mathrm{m\,s^{-1}}$
3. $6.5 \, \mathrm{cm}$

QUESTIONS 1C

2. (a) $2.5 \times 10^4 \, \mathrm{V\,m^{-1}}$ (b) $1.5 \times 10^2 \, \mathrm{V}$

QUESTIONS 3A

1. (a) $1000 \, \mathrm{eV}$ (b) $1.602 \times 10^{-16} \, \mathrm{J}$
   (c) $1.875 \times 10^7 \, \mathrm{m\,s^{-1}}$
2. $8.4 \times 10^6 \, \mathrm{m\,s^{-1}}$
3. $300 \, \mathrm{eV}$
4. $30 \, \mathrm{V}$
5. $1.4 \times 10^7 \, \mathrm{m\,s^{-1}}$

QUESTIONS 3B

1. (a) $4.6 \times 10^{-19} \, \mathrm{J}$ (b) $6.6 \times 10^{-19} \, \mathrm{J}$
2. (a) $4.3 \times 10^{-19} \, \mathrm{J}$ (b) $6.5 \times 10^{14} \, \mathrm{Hz}$
   (c) $4.6 \times 10^{-7} \, \mathrm{m}$
3. (a) $3.4 \times 10^{-19} \, \mathrm{J}$ (b) $2.1 \, \mathrm{eV}$ (c) $2.1 \, \mathrm{V}$
4. $4.0 \, \mathrm{V}$

QUESTIONS 4A

1. $2.4 \times 10^{-10} \, \mathrm{m}$
2. (a) $9.4 \times 10^6 \, \mathrm{m\,s^{-1}}$ (b) $7.7 \times 10^{-11} \, \mathrm{m}$
3. $7.3 \times 10^{-15} \, \mathrm{m}$

QUESTIONS 4B

1. No. (See bottom of p. 138 for explanation.)

QUESTIONS 5A

1. (a) $6.2 \times 10^{14} \, \mathrm{Hz}$, $4.9 \times 10^{-7} \, \mathrm{m}$
   (b) $2.5 \times 10^{15} \, \mathrm{Hz}$, $1.2 \times 10^{-7} \, \mathrm{m}$

QUESTIONS 6A

1. $3.2 \times 10^{-5} \, \mathrm{s}$
2. $3.3 \, \mu\mathrm{s}$
3. $0.94c$
4. $5.1 \times 10^{-5} \, \mathrm{s}$
5. $3 \times 10^{-15} \, \mathrm{s}$

QUESTIONS 6B

1. $11.9 \, \mathrm{m}$
2. $1.8 \times 10^8 \, \mathrm{m\,s^{-1}}$

QUESTIONS 6C

1. (a) $1.56 \times 10^{-5} \, \mathrm{m\,s^{-1}}$ (b) $4.63 \times 10^3 \, \mathrm{m}$
   (c) $6.53 \times 10^2 \, \mathrm{m}$
2. (a) 2.7 light-years   (b) 1.2 light-years   (c) 1.3 years
3. $2.4 \times 10^8 \, \mathrm{m\,s^{-1}}$
4. (a) $3.5 \times 10^{-7} \, \mathrm{s}$ (b) $2.6 \times 10^{-7} \, \mathrm{s}$

QUESTIONS 6D

1. (a) $3.83 \times 10^{-27} \, \mathrm{kg}$ (b) $4.89 \times 10^{-27} \, \mathrm{kg}$
   (c) $8.39 \times 10^{-27} \, \mathrm{kg}$
2. (a) $2.60 \times 10^8 \, \mathrm{m\,s^{-1}}$ (b) $2.98 \times 10^8 \, \mathrm{m\,s^{-1}}$
3. (a) 3.1 (b) 8.8

QUESTIONS 6E

1. (a) $2.5 \times 10^{-30} \, \mathrm{kg}$ (b) $7.1 \times 10^{-22} \, \mathrm{kg\,m\,s^{-1}}$
   (c) $2.3 \times 10^{-13} \, \mathrm{J}$ (d) $8.2 \times 10^{-14} \, \mathrm{J}$ (e) $1.5 \times 10^{-13} \, \mathrm{J}$
2. $2.16 \times 10^{-11} \, \mathrm{J}$
3. (a) $1.64 \times 10^8 \, \mathrm{m\,s^{-1}}$ (b) $1.87 \times 10^8 \, \mathrm{m\,s^{-1}}$
4. $1.32 \, \mathrm{MV}$
5. $1.99 \, \mathrm{MV}$
6. (a) $v = 0.63c$ (b) $M_0 = 2.6m_0$
7. (a) $1.87 \times 10^8 \, \mathrm{m\,s^{-1}}$ (b) $4.20 \times 10^{-11} \, \mathrm{J}$

QUESTIONS 7A

1. $6.8 \times 10^4 \, \mathrm{J}$
2. $3.8 \times 10^5 \, \mathrm{J}$
3. $1.56 \times 10^6 \, \mathrm{J}$

QUESTIONS 7B

1. $158 \, \mathrm{K}$
2. $9.52 \times 10^5 \, \mathrm{Pa}$

INDEX

Parentheses indicate a page where there is a minor reference.

Explanation of answer to Question 4B, 1. The first half of the experiment would produce a single-slit diffraction pattern. The second half would produce a similar pattern slightly offset from the first. The overall effect would simply be the sum of these and would not resemble a genuine double-slit interference pattern.